Published by The Australian Museum,
6-8 College Street, Sydney

ISBN 0 7240 1527 2

Designed by H. & H. Koch, Commercial Artists
Printed by Rodenprint Pty. Ltd.
Typesetting by Tony M. Simmons and Co.

Reprint 1985

Prehistoric Animals of Australia

Based on drawings
by Peter Schouten

Edited by Susan Quirk
and Michael Archer

Prehistoric Animals of Australia is a collection of drawings by Peter Schouten. Many scientists from all over Australia and one from New Zealand have contributed to the text and the Australian Museum is grateful for their assistance. They are as follows:-

Dr. Michael Archer	University of New South Wales
Mr. Ken Aplin	University of New South Wales
Dr. Lyn Dawson	University of New South Wales
Ms. Eileen Finch	University of Western Australia
Mr. Tim Flannery	University of New South Wales
Dr. Ewan Fordyce	University of Otago — New Zealand
Ms. Suzanne Hand	University of New South Wales
Dr. Anne Kemp	University of Queensland
Mr. John Long	Monash University — Victoria
Mr. Mike Plane	Bureau of Mineral Resources — Australian Capital Territory
Mr. Neville Pledge	South Australian Museum
Dr. Pat Rich	Monash University — Victoria
Dr. Tom Rich	National Museum of Victoria
Dr. Alexander Ritchie	The Australian Museum — New South Wales
Dr. Meredith Smith	Institute of Medical and Veterinary Research — South Australia
Dr. Mary Wade	Queensland Museum
Dr. Anne Warren	La Trobe University — Victoria

Contents

Introduction

Michael Archer

Like many of the reasons for the evolution of Australia's unique vertebrates, this volume is the result of pure chance. One day in February, Peter Schouten walked into my office on the off chance that someone would provide some information about the shape of a fossil kangaroo's foot for a drawing he was doing in his spare time. Like everyone else who sees one of Peter's drawings for the first time, all of us in the room suffered instantaneous and severe jaw-drop. His drawing had a feeling of realism that made us wonder if he had used a time machine to photograph the beast. Anyway, with much encouragement, Peter became a regular visitor and soon was born the idea for a joint venture.

Halfway through the drawings, we approached the Director of the Australian Museum (who also suffered severe jaw drop) and Susan Quirk. Susan took on the task of helping to organise appropriate authors for each drawing, of sharing the editorial load of the contributions as they came in and of making publication arrangements.

The result is in your hand — an introduction to 30 of Australia's most remarkable prehistoric citizens.

The Temporal Framework for Australian Vertebrate Evolution

The oldest known vertebrates (from North America) are late Cambrian in age, over 500 million years old. Although I can easily roll that number off my tongue, to me such a time span is almost incomprehensible. To simply count that number of years, taking a mere second to call out each year (which I couldn't do anyway), would take over 95 years of non-stop counting. So it would come as no surprise given 500 million years to play with, that evolution managed to produce the fantastic menagerie of extinct vertebrates that it did.

Geologists have agreed to divide the Earth's history (see the inside cover), into eras (e.g. the Palaeozoic, Mesozoic and Cainozoic Eras), eras into periods (e.g. the Devonian, Cretaceous and Tertiary Periods) and periods into epochs (e.g. the Miocene and Pleistocene Epochs). These conceptual names are part of the international language of palaeontologists and are used to represent the age of the rocks on every part of the surface of the Earth.

How the **actual** age of these rocks is determined is another matter. Ideally, radiometric dating techniques are used to determine the ratio of the amount of a radioactive element to its decay product. This ratio is then used to calculate the length of time that has elapsed since the original radioactive mineral was formed. This very useful technique has enabled, for example, the approximate age of several vertebrate fossil-bearing deposits in Australia to be determined.

Fossil-bearing rocks may be dated indirectly by comparing their fossil remains with those from other rocks that have been radiometrically dated. Many of the rock sequences originally accepted to represent particular periods of time, such as the Cretaceous, were marine. Their kinds of fossil shells, echinoderms, corals and so on were capable of dispersing widely throughout the oceans of their day and can be matched in Cretaceous marine rocks exposed today on almost all continents. It is then only necessary to determine an absolute (i.e. radiometric) date for one of these matched formations in the world in order to have a date that is applicable to them all.

Similarly, assemblages of fossil vertebrates have been used to correlate otherwise undated rocks in Australia. For example, the Chinchilla Sand formation of southeastern Queensland, known since before the turn of the century, contains dozens of different kinds of fossil marsupials. Yet, until recently, there was no way to determine the absolute date of this geological formation. Then, in the 1970s a keen-eyed fisherman from Ayr, Jim Barrett, told the Queensland Museum about fossil bones he had spotted along a creek bank near Charters Towers in northeastern Queensland. After several fossil-collecting expeditions, this Bluff Downs fauna (as it came to be called) proved to be every bit as rich as the Chinchilla fauna and clearly contained similar kinds of fossil kangaroos, koalas and diprotodontids. But more importantly, the Bluff Downs fossils came from sediments that lay under basalt — a datable rock. The age of the basalt turned out to be about 4.5 million years old, which places it in the early Pliocene. After detailed study and comparison of the fossils from the Bluff Downs and Chinchilla formations, it was possible for the first time to give a reasonable estimate of the absolute age of the Chinchilla Sand — i.e., about 4.0 to 4.5 million years or approximately early to middle Pliocene in age.

After many years of this sort of research, we now have a reasonable understanding of the relative and, in some cases, absolute ages of Australia's vertebrate-bearing sediments. The relative positions in time of the fossil vertebrates portrayed in this volume are shown in Table 1. Note that although most species are only known from a limited span of time, some (such as the fossil lungfish **Neoceratodus gregoryi** (p. 22), span a considerable range of time. Clearly, the longer the time range of a fossil species, the less useful it is in correlation.

Naturally there are many other fossil vertebrates known from the Australian record and undoubtedly many more will be found. In the last five years alone, at least 20 new fossil vertebrate sites have been found.

Evolving in a Changing World

Australian vertebrates, like those of every other continent, have been affected by historical events which shaped every aspect of the world in which they lived. Many of the authors of the species accounts following have discussed relevant aspects of the environments in which their subjects lived. This is part of the subject matter of palaeoecology.

Historical factors which were particularly crucial to one group of vertebrates very frequently affected them all, albeit in different ways. For example, when Australia separated from Gondwanaland, the resulting isolation for Australian terrestrial vertebrates was profound. But it also affected the marine faunas because it provided a new southern exchange route between the Indian and Pacific Oceans.

Historical changes which appear to have most profoundly affected the course of evolution for Australian vertebrates fall into three main categories: 1. movements of continents relative to each other; 2. changes in world climates and vegetation; 3. the arrival of Man. We will briefly consider each.

The changes in the position of Australia through time were most profound during the last 45 million years. Although Australia began to separate from Antarctica as early as 90 million years ago it pivoted around a Victorian-Tasmanian-Antarctic connection which did not actually break until sometime between 37 and 46 million years ago. From then on, Australian vertebrates evolved in isolation during which time Australia's distinctive ratite, marsupial and monotreme radiations took place.

Then, about 15 million years ago, the leading edge of the Australian plate began to crumble up against crustal masses in the Indonesian region of southeastern Asia. This had two profound effects. First, New Guinea began to rise as the cool, moist highlands we see today, an event that provided an important habitat when the rest of the Australian continent was drying out. Second, it placed Australia close enough to Asia to receive immigrant groups of Asian animals: bats and monitor lizards (by 15 million years ago), rodents (by 4.5 million years ago), humans (by 50,000 years ago) and dogs (by about 4,000 years ago). These immigrant groups had a profound effect on the descendants of the Gondwanaland groups that arrived with Australia.

Changes in world climates have been continuous throughout the history of vertebrate evolution. In the Devonian, there was a period of considerable drying out. Possibly this was part of the reason that led natural selection to transform some groups of air-breathing fish into forms (the ancestral amphibians) capable of surviving the long overland journeys necessary to find pools that had not evaporated.

Major change came again at the end of the Palaeozoic with the development of extensive glaciers and cold climates. Parts of Australia were covered with ice-sheets. This long episode of cold climates may have been part of the reason for the radiation of the mammal-like reptiles. If some of these could thermoregulate they would have had a distinct selective advantage over their "cold-blooded" neighbours.

Change came again in the Triassic. The world not only emerged from its frigid condition but, in various regions, developed extensive deserts. This reversal of fortunes for ancestral mammals may explain why the "cold-blooded" dinosaurs then began to bloom for the next 100 million years.

Their Waterloo, however, came at the end of the Cretaceous, 65 million years ago. Many things may have happened to cause the biotic crisis that occurred at that time. The angiosperms (flowering plants) were rapidly replacing other forms of plants as the dominant vegetation. Climates were becoming cooler and this would undoubtedly have favoured the mammals. Recent research also suggests that the Earth may have collided at that time with an asteroid about 10 km in diameter. This collision would have wreaked ecological havoc because it would have resulted in a dust cloud that blacked out the sun for several years. Loss of sunlight would have meant collapse of most of the ecosystems of the world. If such an ecological disaster did in fact happen (and the hypothesis is still very controversial), the extinction of dinosaurs, pliosaurs and many other groups at the end of the Cretaceous would be entirely understandable.

The last major climatic change that affected Australian vertebrates was the cooling and drying of the continent which started about 15 million years ago and culminated about 2 million years ago. This was the prelude to the "Ice Ages" of the Pleistocene. With its approach, much of central Australia became arid (progressively more water was being tied up in glaciers), and with this change, a major shift in evolutionary adaption began to take place in Australian vertebrates. Animals that once were adapted to the wet forests of central Australia, were under intense selective pressure to adapt to the spreading woodlands, grasslands and eventually deserts. Many groups that could not adapt fast enough simply vanished. Others, successfully met the challenge and rapidly filled the deserts and grasslands with new and different kinds of animals, those we think of as distinctively Australian such as the Red Kangaroo.

The third sort of major change that affected Australian vertebrates was Man. We have been called "Man the destroyer" by many students of the subject who put in our own laps the responsibility for much if not all of the extinctions that took place in Australia at the end of the Pleistocene. They suggest that the frequent use of fire by Aboriginal Man would have fundamentally changed the nature of all Australian terrestrial ecosystems.

This is another controversial area and not all authorities who have studied the Pleistocene extinctions agree about their cause(s). Others consider that the climatic fluctuations which took place during Pleistocene time were more important as factors leading to extinction. Yet others think that the arrival of Man and climatic change were factors which together led to the extinctions.

Whatever the actual reasons for the Pleistocene extinctions, they should be seen in perspective. In the short space of 150 years, European Man has almost managed to drive more species over the brink of extinction than Aboriginal Man may have managed to do in 50,000 years. Clearly, some of us are much better "destroyers" than others.

In the species accounts following this introduction, the authors have discussed particular aspects of the extinct vertebrates. To help put these selected accounts into perspective, and to point out what seem to be common patterns, the following overview is offered.

Table 1

Era	Period			Selected Australian Vertebrates (or Significant Event)
CAINOZOIC	Quaternary	Recent		Mostly modern species
		Pleistocene		Diprotodon optatum; Palorchestes azael; Meiolania platyceps; Zygomaturus trilobus; Zaglossus ramsayi; Megalania prisca; Thylacoleo carnifex; Troposodon kenti; Wonambi naracoortensis; Bohra paulae; all species of Procoptodon.
	Tertiary	Pliocene	Late	Ischnodon australis; Glaucodon ballaratensis;
			Early	Euryzygoma dunense;
		Miocene	Late	Thylacinus potens; Dromornis stirtoni;
			Middle	Obdurodon insignis; Wabularoo naughtoni; and Megadermatid bat;
			Early	First known Tertiary marsupial discovered at Wynyard, Tasmania
		Oligocene		Oldest known Australian mammals
		Eocene		Australia separated from Antarctica: about 45 million years ago
		Paleocene		Major radiations of mammals throughout the world
MESOZOIC	Cretaceous		Late	Extinction of many groups of vertebrates
			Middle	World's earliest known marsupials
			Early	Muttaburrasaurus langdoni; Kronosaurus queenslandicus; Allosaurus sp. the Qld pterosaur
	Jurassic			Rhoetosaurus brownei; and world's oldest birds
	Triassic			Paracyclotosaurus davidi; and world's oldest mammals
PALEOZOIC	Permian			Major episode of glaciation in Australia and elsewhere
	Carboniferous			World's oldest reptiles
	Devonian			Culmacanthus stewarti; Rolphosteus canningensis; and world's oldest amphibians
	Silurian			Curiously, a total absence of Australian vertebrates from this period
	Ordovician			Oldest known Australian fish from the Northern Territory
	Cambrian			World's oldest vertebrates

Fish: The Oldest Australian Vertebrates

The antiquity of known Australian vertebrates is, by world standards, very definitely respectable. In fact, the recently discovered remains of fish-like vertebrates from Ordovician rocks in the Northern Territory are equivalent in age to the oldest known from anywhere in the world. Their preservation is sufficient to determine that they were simple jawless forms covered in bony plates and scales.

But among fish palaeontologists, Australia is probably better-known for its magnificent record of Devonian fish. From deposits in New South Wales, Victoria and Western Australia have come many strange and fascinating kinds. Some represent genera of fish also known from contemporaneous deposits in Greenland, Europe and even Antarctica. Others appear to be unique to Australia. Much of the more recent work on these Devonian fish has been carried out by Alex Ritchie, John Long, Gavin Young and Roger Miles.

The late Devonian Gogo fauna from northwestern Western Australia is one of the world's most important mainly because of the exquisite preservation of its fish. The Gogo placoderms, armoured fishes with jaws, after death settled quietly into a limey bottom ooze that beautifully preserved their hard parts. Because they were preserved in nodules of what eventually became a very hard protective limestone, their skeletons can now be recovered by etching the limestone with acid. This type of preparation takes months of extremely delicate laboratory work for each of the many fish being prepared.

The Gogo fish are also very diverse. They include placoderms (e.g. **Rolphosteus canningensis**, p. 18), lungfish, primitive ray-finned bony fish and crossopterygians (lobe-finned fish) which once lived in the waters of a fringing barrier reef that surrounded what was then the isolated Kimberley block. The Napier Range is the now high and dry reef which itself was built by the once living algalstranatolite.

The acanthodians (or "spiny sharks" as they are sometimes called although they are not closely related to sharks) are another group of extinct fish with representatives in Australia (e.g. **Culmacanthus stewarti** p. 20) which overlapped the placoderms in time. Most acanthodians are distinguished by the presence of well-developed spines that probably provided stiff leading edges for their fins as well as a pain in the mouth for any carnivorous placoderm that might have been tempted to taste-test one.

There were also many kinds of extinct Australian lungfish, including some very large Cainozoic forms. A large individual of the modern Australian lungfish (**Neoceratodus forsteri**) is certainly an impressive piece of fish, but compared with some of its extinct relatives it seems like small fry. The Eocene to Pleistocene **Neoceratodus gregoryi** (p. 22) grew to about four metres.

As a group, lungfish (rather than the lobe-finned crossopterygians) are now regarded by some students of evolution to be the group basal to the tetrapod radiation. But this speculation is based largely on the similarity in structure between the earliest amphibians (the Devonian labryinthodonts) and lungfish in general. There is, as yet, no known fossil form that is clearly intermediate in structure between these earliest amphibians and any other particular group of fish, at least in the way that **Archaeopteryx lithographica** provides a link between coelurosaurian dinosaurs and birds.

Australian Fossil Amphibians: Setting New Records in Time

Although the origins of amphibians are poorly understood, the Australian fossil record is becoming a good place to look for better evidence. It now appears that late Devonian amphibian tracks in rocks found in Gippsland, Victoria, represent the oldest (albeit by a small margin) known amphibians from anywhere in the world.

The oldest known skeletal remains of amphibians are those of labyrinthodonts from the late Devonian rocks of East Greenland. Now a jaw, discovered in rocks of late Devonian age near

Forbes, New South Wales, has been interpreted to represent a labyrinthodont that also showed features of crossopterygian fish. Clearly, Australia might soon provide the most revealing, as well as the earliest, evidence for the evolution of amphibians.

The Australian record of late Palaeozoic and Triassic labyrinthodonts is rapidly unfolding, in large part through the efforts of Anne Warren, John Cosgriff and their colleagues. Triassic amphibians from the Arcadia Formation of southeastern Queensland, for example, have been found to include, besides a bucketful of novel forms, representatives of the Chigutusauridae, a family previously known only from Argentina.

Labyrinthodonts from Triassic deposits in New South Wales have been known for a considerably longer time. In fact, the two metre **Paracyclotosaurus davidi** (p. 24), from the Wianamatta Group in the Sydney Basin, has been known since 1910. Yet, despite this relatively early discovery, and the fact that it is known from a virtually complete skeleton, there are still major uncertainties about its way of life.

The Australian record of labyrinthodonts also provides the youngest surviving member as well as the earliest known. The "Kolane amphibian", a labyrinthodont which exceeded three metres in length comes from early Jurassic rocks of Queensland. Apart from this, the youngest undoubted labyrinthodonts were middle Triassic in age. This is a good example of a common phenomenon in the Australian fossil vertebrate record — late survival here of a group that elsewhere in the world died out much earlier.

Cainozoic fossil amphibians from Australia are decidely more like living amphibians. The recent do-or-die efforts to find older Tertiary fossil mammals in this country have had fortunate side effects for the amphibian palaeontologists. Mike Tyler has found, among the non-mammal bits in wash concentrates from the middle Miocene sites of central Australia, pelvic bones and other elements representing a fascinating array of extinct frogs. Undoubtedly they represent only a miniscule fraction of the whole record but they are our first glimpse of the fossil record of Australia's frogs.

Australian Fossil Reptiles:
A most Curious Habit of Persistence

The oldest known Australian reptiles (from the Triassic Rewan fauna of Queensland and the Triassic fauna of Tasmania) are considerably younger than the oldest reptiles known — the lizard-like captorhinomorphs from the Carboniferous of North America. But the Australian fossil reptile record is improving at an extremely rapid rate. Most of the recent work has been done by Alan Bartholomai, Mary Wade, Ralph Molnar, Tony Thulborn and Alex Ritchie. Exploration of Mesozoic rocks in Queensland and Victoria has resulted, in the last few years alone, in recognition of Australia's first known pterosaur, a mammal-like reptile and an allosaurid dinosaur.

Because, representatives of many of these groups have been known from other southern continents for a much longer time, their discovery in Australia was not unexpected, just a very long time in coming. That said, there are nevertheless aspects of Australia's fossil reptile record that are very definitely surprising.

Triassic Reptiles

Starting with the Triassic, the rarity of synapsid (mammal-like) reptiles itself is surprising. Elsewhere in the world, including even Antarctica, Triassic terrestrial vertebrate faunas were commonly dominated by these creatures and remained so until the dinosaurs began to diversify. Although one Australian Triassic synapsid has now been found (by Tony Thulborn) a dicynodont evidently similar to species of **Kannemyeria**, it is the only one known despite many years of palaeontological fieldwork in the Triassic sediments of eastern Australia. Why? Perhaps, although Australia was part of Gondwanaland in the Triassic and clearly shared some groups with other continents (such as its labyrinthodonts), it was separated by some kind of ecological filter barrier, perhaps an inhospitable southern climate or an extensive desert.

Jurassic Reptiles

The reptiles of the Australian Jurassic (so far known only from Queensland) are noteworthy for at least two reasons. First, in contrast with most of the Australian Triassic reptiles, all of the Jurassic forms are decidely unlike those from other areas of the world. The sauropod dinosaur **Rhoetosaurus brownei** (p. 26) is apparently one of the most unusual sauropods known. The early Jurassic plesiosaurs (long-necked aquatic reptiles) from Australia may well be the world's oldest freshwater forms.

The second reason why they are noteworthy is the fact that some Jurassic forms seem to represent a persistence of primitive types which died out previously in other areas of the world. **Rhoetosaurus brownei,** for example, although by no means the oldest sauropod, appears to be one of the most primitive. This persistence of relatively primitive forms in Australia is a curiously common feature at most periods in its history, as we shall see.

Cretaceous Reptiles

Of Mesozoic Australian reptiles, the Cretaceous forms are far and away the best known. They have been found in many states but are best represented in Queensland, New South Wales and South Australia. Among the more spectacular is the iguanodontid dinosaur **Muttaburrassaurus langdoni** (p. 30) represented by much of the skeleton of a single individual. Like iguanodontids from other areas of the world, this Australian form had its thumb modified into a spike, possibly for use in defence or in competitions with other individuals for sexual rights. Curiously (because it does not seem to make very good palaeogeographic sense), this Australian form seems to be most closely related to iguanodontids known from the deposits of Central Asia.

There were also less well-represented but more unusual Cretaceous dinosaurs such as a small Queensland ankylosaur that had a series of completely unique bones along its vertebral column and a large sauropod (known from a single neck vertebra) which was evidently in excess of 20 m long.

There is also a species of **Allosaurus** (p. 28) from Victoria which although not particularly unusual per se, is the only Cretaceous allosaur in the world. All others are known from Jurassic sediments — evidently the same story of late survival in Australia of forms which became extinct earlier elsewhere in the world.

While mentioning Australian dinosaurs, note should be made of the remarkable Cretaceous trackway recently discovered near Winton, Queensland. This is one of the largest dinosaur trackways in the world and is providing fascinating information about the social behaviour and mobility of the dinosaurs that stampeded at speeds of 13-15.5 km/hr across the once muddy surface.

But dinosaurs are not the only Cretaceous reptiles known from Australia. Turtles are common in many Cretaceous deposits. Australia's first pterosaur (flying reptile) remains were recently found near Boulia, Queensland, in limestone which had formed in a shallow sea. The sea covered a large area of inland and northwestern Queensland.

Perhaps the most spectacular of Cretaceous reptiles was the marine pliosaur **Kronosaurus queenslandicus** (p. 32). Its skull alone was over two metres long and came complete with a set of teeth which would have encouraged any of even the largest creatures it encountered to stay well out of its way. It is representative of the fact that the Mesozoic oceans teamed with a great variety of reptiles, much in the way that many groups of mammals (e.g. whales, dolphins, dugongs and otters) live in the oceans of today.

As more field work occurs each year, the Australian Cretaceous record in particular improves rapidly. Although it will probably never be as rich as that of some of the northern continents, what it lacks in diversity it will make up for in uniqueness.

Cainozoic Reptiles

The demise of the giant dinosaurs and pliosaurs in Australia at the end of the Cretaceous definitely did not put paid to the evolution of interesting Australian reptiles. Apart from a rapidly growing awareness of a host of ancient but essentially familiar sorts of fossil turtles, skinks, elapid snakes and so on, there were several singularly odd Cainozoic reptiles.

A strange group of crocodiles, the xiphodont crocodiles, which had a deep snout and distinctly shaped shearing teeth, survived here until the Pleistocene. A familiar aspect of this story, however, is the fact that xiphodont crocodiles elsewhere in the world all became extinct by the end of the Miocene. The same late survival story seems to apply to **Wonambi naracoortensis** (p. 40), a very large (five metre) python-like snake from Pleistocene cave sediments in South Australia. Evidently, the closest relatives of this reptile died out elsewhere in the world by the end of the Miocene. Why these peculiar reptiles should have survived for so much longer in Australia is a mystery.

Early Birds

Discovery of Australian fossil birds seems to be attributable to essentially four periods. The first, and for Australian palaeontology as a whole, most significant was the discovery by Mr. Rankin around 1830 of a fossil bird bone in the Wellington Caves of New South Wales. In fact, he had fastened a rope to it thinking it to be a projection of rock. Evidently, he only discovered his mistake after the bone broke under his weight. Discovery of that bird bone marks the real beginning of vertebrate palaeontology in Australia.

The second main important phase of research into Australian fossil birds took place around the turn of the century with discovery of the many fossil birds from Pleistocene deposits exposed in the banks of the inland river systems, in the sediments of Lake Callabonna and in the Pliocene and Pleistocene deposits of the Darling Downs in Queensland. Most of these essentially modern types of birds were named by Charles W. DeVis of the Queensland Museum.

The third major phase permitted the first significant look at Australia's middle Tertiary birds. This came with the pioneering work of Professor Ruben Stirton and his colleagues in the 1950s and 1960s. In particular, Alder Miller described at this time several Tertiary birds, including flamingoes, pelicans and emus.

But it was during the fourth and present phase, which followed tight on the heels of the third, that Australian bird palaeontology really began to take off. Largely through the efforts of Pat Rich and Gerry vanTets, a great variety of new, as well as previously collected, but little-studied fossil birds, have been described.

Pat Rich's monograph of the Dromornithidae is an important milestone in understanding this fascinating group of extinct Australian birds. All were large, flightless, herbivorous ratites similar to, but different from, emus and cassowaries. And among them, the late Miocene **Dromornis stirtoni** (p. 42) stands out, not only as the largest dromornithid, but quite probably as the largest bird that ever lived anywhere in the world.

As part of the world radiation of flightless ratites, these extinct Australian dromornithids (as well as the family of emus and cassowaries) are becoming a focus of much interest to palaeobiogeographers. Their complete disappearance at the end of the Pleistocene, along with many kinds of large Australian mammals, has focused attention on the mysterious causes for the late Pleistocene extinctions (see below).

A final note about Australia's venerable birds rests on feathers. From early Cretaceous lake sediments in Gippsland, Victoria, have come a variety of small feathers. They are clearly bird feathers, although it is not clear what sort of bird they represent. Considering that this record is not much younger than the Jurassic age of the oldest known bird **Archaeopteryx lithographica,** birds must have reached Australia soon after they evolved.

Mammal Beginnings in Australia: A Maddening Puzzle

Australia's oldest known mammals are unfortunately, rather young. From other areas of the world, notably Europe, mammals are known to have been in existence for over 200 million years, i.e. since at least late Triassic time. Australia at that time was still an integral part of Pangaea (the ancient supercontinent) and so ought to have been in the running for a share of the world's oldest mammals. Yet, so far the oldest Australian mammal fossils are only about 23 million years old.

This situation of ignorance of the older record is maddening and hopefully will end soon. Three things give hope that new finds will soon advance the frontier of early Australian mammal history. The first is the recent discovery of mammal-like reptiles from Queensland, a fact which suggests that the ancestors of mammals were here (as well as elsewhere). The second is the recent discovery of Eocene mammals in Antarctica, a fact which suggest mammals were widespread across Gondwanaland at least as early as 45 million years ago. The third reason is a bit less direct. Discovery of fossil fleas from late Cretaceous sediments in Victoria (the same sediments containing the feathers mentioned earlier) has led an entomologist to declare that the mouthparts of one of the fleas are of the kind found today only in mammal-biting fleas. So, now all we evidently have to do is find its Cretaceous victim.

Efforts to find the earlier mammals of the Australian "dark ages" are going on every year. Exploration of Mesozoic rocks has, as noted above, yielded many new and exciting finds such as dinosaurs, pterosaurs and mammal-like reptiles. Surely, it will only be a matter of time before a tiny mammal jaw will pop out from the toes of a dinosaur. Similarly, continuous searches in early Tertiary rocks in eastern Australia should, before long, turn up Paleocene or Eocene beasts.

And what will they look like? Well, that is part of the fun of the search — we don't really know what to expect. However, it is reasonable to guess that there will be primitive marsupials (pouched mammals) as ancestors for the later Tertiary to modern marsupials which we **do** know about. What **sort** of primitive marsupials would there have been? Presumably some would have closely resembled kinds of early Tertiary marsupials known from South America. But which of the many kinds known from South America? Others might be early monotremes (egg-laying mammals) as ancestors for the platypuses and echidnas and, perhaps, bats. Bats are **known** as fossils in Australia at least as early as 15 million years ago. Clearly, these are many fascinating questions but, without those older Australian mammals, we have no answers.

Mammals of the Later Cainozoic

The middle to late Tertiary mammals of Australia are becoming reasonably well known. They included many forms which, although referable to still living families, were clearly more primitive than living forms. For example the middle Miocene **Obdurodon insignis** (p. 44) was clearly a kind of platypus but one much less specialised than the living species. The middle Miocene **Wabularoo naughtoni** (p. 66) was similarly an undoubted kangaroo but one representing a subfamily which has no living representative. The Pliocene **Ischnodon australis** although clearly a rabbit-eared bandicoot, was more like other kinds of bandicoots than its living relative (**Macrotis lagotis**). The middle Miocene false vampire bat from Riversleigh is clearly related to the living Ghost Bat (**Macroderma gigas**) but is much smaller and differs in various aspects of dental and skeletal anatomy.

Other Tertiary mammals represent families which have no living descendants. The middle Miocene species of **Neohelos** (p. 56) were diprotodontids, a family which included dozens of kinds of large herbivorous (plant-eating) marsupials which were completely replaced by the explosive radiation of kangaroos. But before finally passing into stone, the diprotodontids produced several extraordinary forms. As examples, there were the bizarre early Pliocene **Euryzygoma dunense** (p. 60), the knobbly-skulled Pleistocene **Zygomaturus trilobus** (p. 58), and the largest marsupial ever, the Pleistocene **Diprotodon optatum** (p. 62).

Among carnivorous (meat-eating) mammals of the Tertiary, there were also some remarkable productions. Discovery of a dolphin from the Miocene sediments of the Frome Basin was one of the biggest surprises of recent years. There are no living freshwater dolphins on this continent although there are in other areas of the world. The large carnivorous marsupial **Thylacinus potens** (p. 48), a Miocene relative of the living Tasmanian "Wolf", must have terrorized its share of kangaroos as would have **Glaucodon ballaratensis** (p. 50), a powerfully built Pliocene dasyurid.

Australian mammals of the Pleistocene included a suite of giants called collectively the "megafauna". The herbivorous **Palorchestes azael** (p. 54), a Pleistocene palorchestid, was probably the most bizarre of these giants. With its trunk, claws and powerful forearms, it represents a completely unique sort of mammal. The Pleistocene species of **Procoptodon** (p. 72) were giant short-faced, single-toed kangaroos. **Troposodon** species (p. 70), although less bizarre, were just as interesting. They appear to have been relatives of the species of **Procoptodon** but perhaps even closer, albeit giant, relatives of the living Banded Hare Wallaby (**Lagostrophus fasciatus**).

Other Pleistocene giants are also noteworthy because of their surprising distributions. **Bohra paulae** (p. 68), the giant tree-kangaroo from the Wellington Caves of New South Wales, represents a type of kangaroo not previously known to occur anywhere south of northeastern Queensland. Species of **Zaglossus** (p. 46), the giant long-beaked echidnas, are today confined to New Guinea. Pliocene and Pleistocene long-beaked echnidas from southern Australia indicate a formerly much wider distribution for these enigmatic egg-layers.

Of all Pleistocene marsupials, probably the most fascinating is the Marsupial "Lion", **Thylacoleo carnifex** (p. 52). Its meat-eating abilities, probable tree-living habits and leopard-size conjure up a most extraordinary image. And its almost certain overlap with Aboriginals would have provided the first Australian people with a very real reason to fear the night.

Twilight of the Giants

Between about 30,000 to 20,000 years ago most of Australia's "megafauna" vanished. The reasons why **Diprotodon optatum, Megalania prisca, Meiolania platyceps,** dromornithid birds, **Thylacoleo carnifex** and many other large species became extinct at the end of the Pleistocene is almost as mysterious and controversial as the reasons for why the dinosaurs died out. However, two basically distinct sorts of reasons have been proposed: the ultimate effects of climatic change and the effects of Aboriginal man.

Certainly climatic changes towards the end of the Pleistocene were profound. At the peak of the last eposide of glaciation, precipitation across Australia evidently reached an all-time low. Even the small areas of montane rainforest in northeastern Queensland all but disappeared. The overall effects of this drop in rainfall on the vegetation upon which the megafaunal species depended evidently resulted in an intense drought, lasting at least several thousand years. Stresses of this sort on populations of large animals can be critical.

On the other hand, Aboriginal Man would undoubtedly have exerted at least as severe, although perhaps more localized, effects. Studies of pollen-rich sediments deposited in Lake George, suggest that there was a marked increase in charcoal about 120,000 years ago, a feature interpreted by some research workers to indicate the time of arrival of Aboriginal Man into Australia with his fire-stick — the increase in charcoal dust being the trademark of human activity. This increase in burning would have resulted in, among other things, a spread of grasslands, a reduction of rainforests in favour of fire-adapted schlerophyll forests and so on — changes in habitat that could well have tipped the fine balance of survival for the larger mammals.

We may never know the precise reasons for the extinction of the magnificent "megafauna", but undoubtedly a lot of factors were involved, and perhaps not the same ones for each species lost. Further, not all large species vanished. The living Eastern and Western Grey Kangaroos, Red Kangaroos, Emus, antilopine Kangaroos and Wallaroos are very large animals, larger than many species of the "megafauna" which vanished 30,000 years ago.

About the artist

Peter Schouten is the artist who has brought the animals of prehistoric Australia to life in this book. He was born and grew up in Petersham, an inner suburb of Sydney. He was educated at Christian Brothers High School at Lewisham and later at Sydney Technical College where he gained the Architectural Drafting Certificate.

His extraordinary ability in depicting the animals on these pages is not only inherited from his mother and father, but has also been generated by a hunger for the nature lacking in his home suburb. In his own words: "my isolation from bushland reinforced my appreciation for natural history because it possessed an element of mystery, irresistible to a small boy. That same element of mystery probably accounted for my keen interest in drawing prehistoric life. To me these were entire worlds of unfamiliar creatures I could explore with my imagination and a pencil." Below Peter describes how he brought the animals in this book back to life sometimes from the evidence of only a few bones.

"The job of reconstructing prehistoric animals must begin with the work done by palaeontologists, both in the field and in the laboratory. The fossils must first be discovered, restored and then studied. Without the dedication of these people my work would be impossible.

"To arrive at a finished illustration I must complete three important stages:

• "The first is where I try to learn as much about the subject as possible. Here the palaeontologists are essential as they provide me with the basic information from which the animal is constructed such as: the age of the fossil and how comprehensive its remains are; habitat at the time of the living animal; habits of the animal deduced from dentition and skeletal structure and the known or hypothetical affinities, both generic and analogous, with other contemporary and prehistoric animals.

• "The second stage is reconstruction. I gather information which is of particular relevance to my needs for rebuilding the animal. Construction starts with the bones. This is easy when there are essentially complete remains. However, the fossils are usually incomplete and quite often fragmentary, and therefore I need the guidance of the palaeontologist to piece together the complete skeleton. I then add the muscles to the skeleton using comparisons with closely related species or with those having a similar lifestyle. I can decide on the outward appearance of the animal. Was it furred, feathered, scaly or did it have a thick tuberculated hide? What colour pattern did it have? These factors are dependent on environment, habits and sexual dimorphism and are largely based on studies of contemporary animals of similar lifestyle.

• "The third stage is animation. It is here that I give the reconstructed animal motion by depicting it in a life-like posture in a believable setting. To accomplish this I need to study the movement and posturing of similar animals during activity and at rest. I make some sketch studies and decide on a particular position according to its information and aesthetic values. The animal is placed in a suitable habitat and the final drawing is then rendered."

Peter's work has been referred to the scientists who have contributed to this book and all agree that his reconstructions are not only as correct as they can assess, but they are also truly amazing.

① RECONSTRUCTION OF SKELETON
② REBUILD MUSCULATURE
③ COMPLETE ANIMAL WITH FUR COVERING SHOWING DIRECTION OF FACIAL GROWTH
④ POSTURING

①

②

③

④

P. M SCHOUTEN '83

17

Armoured tube-nosed fish
Rolfosteus canningensis

John Long

The placoderms were a diverse group of shark-like fishes which flourished during the Devonian period and were extinct by the Carboniferous. The head and trunk of these fishes were covered by a mosaic of boney plates which are often the only fossilized remains of the fish. Some of the best material of placoderms in the world has come from near Gogo station in northwestern Australia. These fossils occur in limestone nodules which can be dissolved in acetic acid to free the bone. This makes it possible to attain perfect three dimensional restorations of the placoderm armour (most of the Gogo placoderms belong to a specialized group known as "euarthrodires" (true jointed neck)). Euarthrodires possessed two pairs of upper jaw toothplates and a well-developed neck joint which articulates the head and trunk shields. Perhaps one of the most bizarre forms from this fauna was **Rolfosteus canningensis,** an euarthrodire with an elongate, tubular snout. **Rolfosteus canningensis** has a streamlined body for fast swimming, large eyes and specialized snout for sensing food and predators and relatively weak jaws for catching prey. It is difficult to imagine what kind of lifestyle **Rolfosteus** led but probably it fed on smaller fishes or invertebrates which lived in the ancient reef environment. **Tubonasus species,** a similar form to **Rolfosteus** with a tubular snout, possessed crushing jaws and probably fed on crustaceans.

Occurring with **Rolfesteus canningensis** and **Tubonasus** in the Gogo fauna are several other euarthrodires. They had powerful jaws for catching other fishes. A small placoderm that had prop-like boney arms, **Bothriolepis,** and a chimaerid-like placoderm, **Ctenurella,** are representatives of the fauna which also occur in the northern hemisphere. Besides placoderms, bony fishes such as lungfishes and ray-finned fishes (or palaeoniscids) are common elements in the fauna. (Long, 1982.)

CLASS: Elasmobranchiomorphi
SUBCLASS: Placodermi
ORDER: Euarthrodira
FAMILY: Camuropiscidae
GENUS AND SPECIES:
Rolfosteus canningensis. Dennis & Miles, 1979b.

Spiny deep-bodied fish
Culmacanthus stewarti

John Long

Acanthodians were a group of primitive fishes which inhabited both marine and freshwater environments from the Silurian period to the Permian period. They are characterized by possessing bony fin-spines at the front of each fin, and by their minute scales. These parts are often the only fossil remains of the fish as entire acanthodians are rare. The recent discovery of entire, well preserved acanthodians from Mt. Howitt, central Victoria, has provided important information on Australian acanthodians. Amongst these was a peculiar deep-bodies form called **Culmacanthus stewarti**.

Culmacanthus (from the aboriginal word "culma", a spiny fish) is known only from a few specimens, one of which is the entire body in lateral view. The head is typified by large plates of bone covering the cheeks with small bony scales or tesserae covering the rest of the head. The shoulder girdle has three plates on the underside which are not rigidly connected to the pectoral fin-spines. This separates **Culmacanthus** from other primitive deep-bodies acanthodians. The massive spines were primarily for warding off attackers as **Culmacanthus** was a browsing fish and was not suited to fast straight line swimming to escape predators. The site where Culmacanthus was found represents an ancient lake deposit, so it is not unreasonable to imagine this fish nibbling at the algae or weeds growing on the lake bottom in the shallows. If pursued by a larger fish **Culmacanthus** could erect its fin-spines or scoot quickly into the cover of the weeds.

Culmacanthus is found associated with slender acanthodians, amoured shark-like fishes (placoderms) and a variety of bony fishes (osteichthyans) from the Late Devonian Mt. Howitt locality. Cheek plates and spines from a similar acanthodian to **Culmacanthus** have also been found from Freestone Creek, north of Briagolong, Victoria. Elsewhere in Australia acanthodians are known from fragments only.

CLASS: Teleostomi
SUBCLASS: Acanthodii
FAMILY: Culmacanthidae
GENUS AND SPECIES:
Culmacanthus stewarti Long 1983b.

A giant lungfish
Neoceratodus gregoryi

Anne Kemp

The illustration shows a reconstructed scene from the Miocene of central Australia. The giant lungfish, **Neoceratodus gregoryi,** slips silently from the cover of the lily pads, apparently to pounce on an unsuspecting duck. Luckily for the ducks, but less exciting for us, this is probably not what it was going to do. More likely, the cumbersome fish was rising through the lily stems simply to exhale noisily among the leaves at the surface, and to refill its lungs with fresh air. The startled ducks would fly off and the fish would drop back to the bottom of the lake. There it would have remained for hours or even days before surfacing again.

This giant lungfish was probably capable of lung and gill breathing like its modern Australian relative, **Neoceratodus forsteri,** and could possibly have survived for some time in cool damp mud and plant debris if the water levels in the lake fell. Among lungfish in general, only the living African and South American are obligate air breathers, having to surface for air regularly.

Fossil remains of the teeth and bones of **Neoceratodus gregoryi** are sufficiently well-known to enable reconstruction of the animal. All lungfish belong to the subclass Sarcopterygii, or lobe finned fishes, along with the living coelacanth and many fossil forms, but they are placed in their own order, the Dipnoi. Although lungfish have been around since Devonian times (350 million years ago), **Neoceratodus gregoryi** is a comparative newcomer, having been found in Eocene deposits in Queensland (38-26 million years ago) and in Miocene and Pleistocene deposits in South Australia (26-5 million years ago and 1.25 million years ago respectively).

The giant lungfish lived in the large freshwater lakes that once covered much of central Australia. Their diet, to judge from the dentition, consisted of quantities of water plants and small aquatic animals like snails. These fish attained a large size. The proportions of the teeth and skull bones which have been found suggest that they grew as big as 4 m in length.

Like the living **Neoceratodus forsteri,** the skull was made up of cartilage surrounded by several large bony plates for additional protection. The dentition consisted of two large crushing plates in the upper and lower jaw. The lips were fleshy, without supporting bone. The body was probably covered with large scales which lay beneath a smooth slimy skin.

CLASS: Teleostomi
SUBCLASS: Sarcopterygii
ORDER: Dipnoi
FAMILY: Ceratodontidae
GENUS AND SPECIES:
Neoceratodus gregoryi White, 1925.

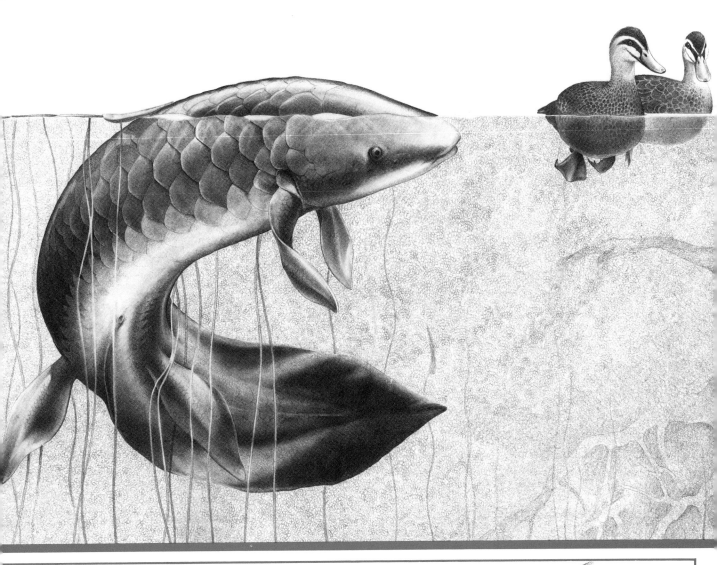

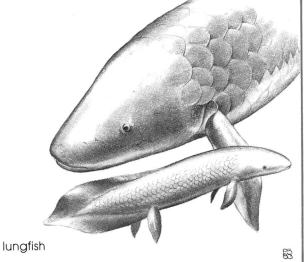

Compared with modern lungfish

An enormous amphibian from the Sydney Basin
Paracyclotosaurus davidi

Anne Warren

Late last century, Mr B. Dunstan of the Geological Survey of New South Wales began collecting fossil vertebrates from Middle Triassic lake deposits exposed in a series of brick pits at St. Peters, New South Wales. Dunstan recovered numerous specimens of fish which were described by Smith Woodward and R. T. Wade, and one giant labyrinthodont amphibian, **Paracyclotosaurus davidi**.

This single specimen of **Paracyclotosaurus davidi** remains one of the most complete labyrinthodont amphibian skeletons known. It is also one of the largest, measuring over 2.25m from the front of the skull to the tail tip. Yet, although it was extremely big compared with its contemporaries **P. davidi** is in other ways typical of labyrinthodonts of the Triassic.

Its head was large and the skull was covered with **ornament** into which was impressed a series of sensory canals. These can be compared in function with those of the lateral lines of modern fish. Its orbits were situated far back on the skull and towards the midline so that unless the eyes could be projected above the head, as they can in modern frogs, the animal's only clear view would have been upwards. The limbs are not well-developed and, although they would have been capable of taking the animal on short trips ashore, **Paracyclotosaurus davidi** undoubtedly led a primarily aquatic life.

Scaly skin impressions found with **Paracyclotosaurus davidi** suggest that the animal's skin was dry so that, unlike the modern amphibia, it may not have used its skin to help gas exchange in breathing. Instead, the expanded thoracic ribs indicate that **P. davidi** had developed a lung ventilating mechanism like that of reptiles.

Watson (1958) suggested that its body shape, weak teeth, largely cartilagi nous skeleton and dorsally placed eyes were indications that **P. davidi** was rather like a giant salamander. It may have hunted by laying in wait, presumably on the lake bottom, until prey came near. Then, by quickly opening its mouth, it could suck in its prey, probably holostean fish such as the 15cm long species of **Promecosomina**.

Chernin and Cruikshank (1978), however, likened the body shape and cross section of **Paracyclotosaurus davidi** to that of a shark, and suggested that its way of life could have been more crocodilian. In their view ". . . the animal could have lurked in neutral buoyancy with eyes protruding just above the water as in a modern crocodile. Prey approaching the water's edge would be taken by a sudden lunge and dragged into deeper water . . .". It is most likely that the true picture lies somewhere between these two, and it is unlikely that it will ever be known with certainty.

Capitosaurids, members of the family to which **Paracyclotosaurus davidi** belongs, had a wide Australian distribution in the Triassic being known from sediments in New South Wales, Queensland, Western Australia and Tasmania.

CLASS: Amphibia
SUBCLASS: Labyrinthodontia
ORDER: Temnospondyli
FAMILY: Capitosauridae
GENUS AND SPECIES:
Paracyclotosaurus davidi Watson, 1958

Australia's oldest giant sauropod dinosaur
Rhoetosaurus brownei

Mary Wade

The very word dinosaur flashes up the picture of the four-footed, long-necked, long-tailed reptiles known as sauropods. Too heavy to run, they walked with head held high like a giraffe in order to carry the weight of the neck on bone-to-bone contacts. All the sauropods were plant-eaters and had narrow teeth projecting 1-6 cm above the gums. Reptiles replace teeth throughout life, so these were adequate for browsing on conifer trees, seed ferns and ferns. There were no grasses in Mesozoic time.

One of these, **Rhoetosaurus brownei,** was found near Roma, Queensland, in 1924. Half the tail, a complete hind leg, some body vertebrae and ribs, and one and a half long neck vertebrae were found, besides a mass of fragments. The restored animal was about 17 m long, just over 3 m high at the hip, and of average bulk (perhaps 20 tonnes in weight). The remains were buried in a large, sandy outwash plain 170-180 million years ago (basal Middle Jurassic time).

World climate at the time was mild. Although the area around Roma was then about 50° latitude, it was subtropical to warm temperate. At these temperatures the problem of a large, thick-bodied animal is not how to warm up, but how to keep cool. One calculation indicated that it would have taken about two days to air-cool a giant sauropod a few degrees without using sweat (reptiles do not sweat). Extra internal air-passages and blood flow to the surface could have helped, but if half of each day was sunny, a 400% increase in the supposed cooling efficiency would have been required to cool a sauropod that walked about in the summer sun. Perhaps they used the same technique as the hippopotami, and slept the hot hours away under water, going out to browse only during the late afternoon and night.

Calculations of dinosaurs' speeds have been made from the measurements of their limbs. **Rhoetosaurus brownei** could have walked at any speed up to 15 km/hour. Browsing and moving at about 2 km/hour, it could easily have covered 20-30 km in a night's travel.

CLASS: Reptilia
ORDER: Saurischia
FAMILY: Incertae sedis
GENUS AND SPECIES:
Rhoetosaurus brownei Longman, 1926

Relative to size of man

Cretaceous killer extraordinaire

Allosaurus species

Tim Flannery

The various kinds of **Allosaurus** were the largest and most impressive carnivores of their day. Up to 10 metres long and nearly 4 metres high, the allosaurs could probably prey on dinosaurs as large as **Apatosaurus** (formerly known as **Brontosaurus**). Despite the vast bulk of these allosaurs, they were not very bright because they possessed a brain about the size of a cat's. The name **Allosaurus** means different lizard. They received this uninspiring name simply because their obviously impressive remains were different from those of any of the other dinosaurs known at the time.

During the Jurassic and Cretaceous periods, when the allosaurs existed, most of the world's landmasses were connected and the climate was much more uniform than today. These factors allowed many kinds of plant and animals such as the allosaurs to have a world wide distribution.

The Australian allosaur is represented by a single ankle bone. The bone was found preserved along with the remains of smaller herbivorous dinosaurs in sandstones which outcrop as coastal cliffs and rock platforms near Inverloch, Victoria. The ankle bone of the Victorian allosaur is more sturdily built than those of other allosaurs, and suggests that this species was more robust.

The Victorian species is important because it is the latest surviving allosaur known. Elsewhere, allosaurs became extinct by the end of the Jurassic period 135 million years ago. The Victorian species is about 125 million years old or early Cretaceous in age.

Interestingly, when the Victorian allosaur stalked the Earth, southern Victoria was well within the Antarctic Circle. But there were no polar ice caps, and plants and animals flourished even at these high latitudes. Apart from the allosaur, other relict species are known from the early Cretaceous rocks of Victoria, and it is possible that the south polar region acted as a refuge for species which became extinct earlier elsewhere.

CLASS: Reptilia
ORDER: Saurischia
FAMILY: Allosauridae
GENUS: **Allosaurus;**
the species is yet to be determined.

Killing juvenile sauropod

Bird-footed dinosaur from Queensland

Muttaburrasaurus langdoni

Alexander Ritchie

Dinosaur discoveries in Australia are few and far between and most of them are disappointingly incomplete. As more finds come to light and the search intensifies, the situation is slowly changing and exciting finds undoubtedly await discovery.

The most complete dinosaur skeleton yet discovered in Australia is that of **Muttaburrasaurus langdoni** which comes from early Cretaceous marine deposits of central Queensland. It consists of a skull and most of the skeleton but lacks much of the tail. The fossil bones of **M. langdoni** came to light in a cattle mustering area on the banks of the Thomson River, near Muttaburra, 100 km. NNE of Longreach, Queensland. Before the significance was recognised, many of the bones had been broken and scattered by the hooves of cattle and others had been 'souvenired' by local residents. In 1963, after the find was reported to the Queensland Museum by a local grazier, Mr. D. Langdon, the Director, Mr. (now Dr.) Alan Bartholomai and his staff visited the site and recovered what was left. A public appeal led to the return of most of the parts which had been removed by locals. After years of painstaking work it became obvious that they came from a completely new type of dinosaur belonging to a group, the iguanodontids, which were well-known from finds in the northern hemisphere.

Muttaburrasaurus langdoni Bartholomai and Molnar (1981) was about 7 m long, able to walk on its hind legs, but probably spending much of its time browsing on all fours. It is an ornithischian ("bird-hipped") dinosaur belonging to the suborder Ornithopoda ("bird-foot"). In many respects it is most closely related to the family Iguanodontidae, which includes **Iguanodon** and **Camptosaurus**, representatives of which have been found on every continent except South America and Antarctica.

One of the most distinctive features of **Muttaburrasaurus langdoni** is an inflated, hollow bony roof over the snout in front of the eyes, the function of which is not known. The front of the mouth lacked teeth. It was probably developed as a horny beak, like most ornithopods. The rows of cheek teeth differed from those of most other ornithopods. Because of this, Bartholomai and Molnar (1981) have suggested that although **M. langdoni** was mainly a plant eater, it may also have been partly carnivorous.

We know little about the environment in which **Muttaburrasaurus langdoni** lived. Although it was unquestionably a terrestrial animal, its skeleton was buried in shallow-water marine sediments with a rich assemblage of molluscs (bivalves, gastropods and occasional ammonites). This suggests that the carcase was washed out to sea after death, sank and was buried with the marine invertebrates. When the work of restoring the skeleton has been completed **Muttaburrasaurus** will go on permanent display in the newly built Queensland Museum in Brisbane.

CLASS: Reptilia
ORDER: Ornithischia
FAMILY: Iguanodontidae
GENUS AND SPECIES:
Muttaburrasaurus langdoni Bartholomai and Molnar, 1981.

Giant killer of the Cretaceous seas
Kronosaurus queenslandicus

Alexander Ritchie

During the Mesozoic Era, 230-63 million years ago, a spectacular array of large reptiles, mainly dinosaurs, dominated the Earth's land masses. The Mesozoic oceans were also inhabited by various groups of large, non-dinosaurian reptiles such as ichthyosaurs, plesiosaurs (and their relatives the pliosaurs) and mosasaurs, all of which died out about the same time as the dinosaurs.

Kronosaurus queenslandicus, the largest known marine reptile, lived around 100 million years ago, in early Cretaceous times, when shallow seas covered much of central Australia. It was first described in 1924 from a fragment of lower jaw discovered near Hughenden, Queensland, in 1899. Heber Longman, then Director of the Queensland Museum, identified it as part of a pliosaur, an extinct, long-skulled, short-necked marine reptile, a distant relative of the better-known, small-skulled and long-necked plesiosaurs. Longman named it **Kronosaurus queenslandicus,** referring to the predatory nature of its large, conical teeth. The Greek god Kronos was reported to have killed and eaten his children to prevent them from usurping his throne.

In 1931-2, W. E. Schevill from the Museum of Comparative Zoology in Harvard University, visited Australia in search of fossils and, from sites north of Richmond in northern Queensland, collected two **Kronosaurus queenslandicus** specimens — a partial juvenile skull and most of the skull and skeleton of a large adult individual. These were shipped back to the United States, prepared over a period of some 25 years, and finally went on display as one of the prize exhibits in the Museum of Comparative Zoology in the late 1950's. Missing portions of the skeleton were restored using specimens of other kinds of pliosaurs as a guide.

In 1935, a partial skull was found between Richmond and Hughenden and, in 1979, the Director of the Queensland Museum, Alan Bartholomai, collected another skull and much of the associated postcranial skeleton of a **Kronosaurus queenslandicus** specimen which is currently being prepared for eventual display in Brisbane.

Although **Kronosaurus queenslandicus** may have been exceeded in length by some of the more lightly built, long-necked plesiosaurs, it remains one of the bulkiest and largest known pliosaurs, a terror of the ancient seas that preyed on fish, turtles, ichthyosaurs and plesiosaurs. It reached around 14 metres in length of which about 25% was taken up by the massive skull with its formidable array of powerful conical teeth.

CLASS: Reptilia
ORDER: Sauropterygia
FAMILY: Pliosauridae
GENUS AND SPECIES:
Kronosaurus queenslandicus Longman, 1924.

Relative to size of man

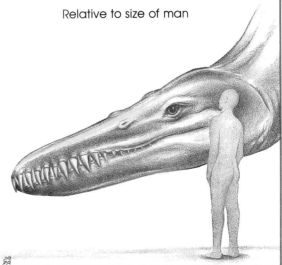

First discoveries of pterosaurs in Australia

Alexander Ritchie

During the Age of Reptiles, when dinosaurs dominated the land and ichthyosaurs and plesiosaurs ruled the seas, a different but equally remarkable group of reptiles took to the air: the pterosaurs. Literally meaning "winged lizards", pterosaurs were a diverse and very successful group between late Triassic times (about 190 million years ago) until the end of Cretaceous times (65 million years ago) when most of the ruling reptiles became extinct. Pterosaurs ranged in size from the dimensions of a sparrow to the largest flying animals the world has seen, remarkable creatures with a wingspan around 15 m. Although they were very delicately built animals and difficult to preserve as fossils, their fossilized remains have been recovered from Mesozoic rocks on all continents with the exception of Antarctica and, until recently, Australia.

Many scientists suspected that pterosaurs once existed in Australia because the dinosaurs, their more restricted, land-bound contemporaries, managed to reach and diversify on this continent between 200 and 65 million years ago. It would have been strange if, with the ease of movement through flight, the pterosaurs hadn't managed to reach Australia when the dinosaurs did.

But it was not until 1979 that the first Australian pterosaur fossils were actually found. They were discovered at a site east of Boulia in western Queensland, in early Cretaceous marine limestones. The bones were carefully extracted by dissolving the surrounding limestone with acid. Unlike most pterosaur fossils found overseas, which are usually preserved flattened and crushed, the Queensland ones, although extremely delicate, are virtually uncrushed and undistorted. Only a few bones have been recovered to date including parts of a lower jaw, a vertebra and a scapulocoracoid (part of the shoulder girdle), but these are sufficient to establish the presence of pterosaurs here. From the bones so far found it seems that this first, as yet unnamed, Australian pterosaur was a moderately small, sea-going fish-eater with a wingspan of about 2 m. The jaw fragment has widely spaced sockets for small, conical, slightly flattened teeth not unlike those of **Ornithocheirus,** a pterosaur from the early Upper Cretaceous of England. The scapulocoracoid, however, closely resembles that of late Cretaceous species of **Pteranodon.** The restoration shown, based on this meagre material, is obviously tentative until such time as more of the extensive outcrops of Cretaceous rocks east of Boulia might provide the answers and reveal precisely how these Australian pterosaurs are related to better-known forms from deposits overseas.

CLASS: Reptilia
ORDER: Pterosauria
FAMILY: ?Pteranodontidae
GENUS AND SPECIES: not yet described.

Conjectural posture, feeding young

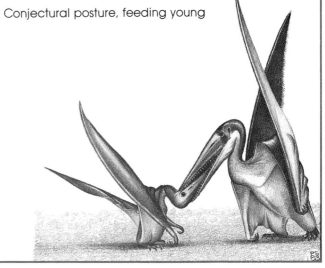

A horned turtle from Lord Howe Island
Meiolania platyceps

Alexander Ritchie

Lorde Howe Island, an isolated, eroded volcano in the Tasman Sea 644km, ENE of Sydney, New South Wales, is a mecca for tourists with its sandy beaches, coral reefs, palm trees and two spectacular mountain peaks rising almost sheer from the sea. Lord Howe is also attractive for scientists because of its fascinating range of plants and animals, many unique to the island. But the most unusual inhabitant, unfortunately long extinct, was the large, land-living turtle, **Meiolania platyceps,** whose fossilised bones have been recovered in considerable numbers from 40 to 50,000 year old coral-sand deposits near the north end of the island over the past century. Although they were originally interpreted by Sir Richard Owen in 1886 as the remains of a giant horned lizard, the discovery of better preserved specimens, including two skulls, proved they belonged to a very strange, extinct turtle.

Looking like something out of a science fiction story, **Meiolania platyceps** was almost 2 m in length with sharp cow-like horns on its head and a spiked club on its tail. As fossil turtles in general go, this one is extremely well-known. There are now hundreds of individual bones, about half a dozen good skulls, several incomplete shells and numerous tail-sheaths. And although no complete skeleton has yet been recovered, available material, mostly in the collections of the Australian Museum in Sydney, has enabled one of the world's leading authorities on fossil and living turtles, Dr Eugene Gaffney of the American Museum of Natural History in New York, to construct a composite replica of a whole skeleton of **M. platyceps.**

Its shell (or carapace) is not especially unusual but its large, triangular skull with the pair of curved, pointed horns at the rear, is unique among turtles. **Meiolania platyceps** obviously could not have retracted its head into the shell but, at least on Lord Howe Island, where there were few if any predators large enough to worry it, it probably didn't need to. Its short stumpy legs, with powerful clawed toes, were clearly adapted for life on land, not for swimming. The relatively long tail was covered with articulating bony rings and ended in a long, conical bony sheath armoured with pairs of spikes.

Meiolania platyceps belongs to a group of turtles, the meiolaniids, that have a long history. The oldest representatives of this group are known from 60 million year old rocks in Patagonia, South America. Other meiolaniids are known from Pleistocene sediments in areas of eastern Australia. How the last surviving member, **Meiolania platyceps,** first reached Lord Howe Island, how long it survived there, why it became extinct in the apparent absence of obvious predators, and what it used its horns and spiked tail for are most curious but unsolved problems.

CLASS: Reptilia
ORDER: Testudines
FAMILY: Meiolaniidae
GENUS AND SPECIES:
Meiolania platyceps Owen, 1886.

The ancient giant butcher
Megalania prisca

Thomas H. V. Rich

Megalania prisca, which means the "ancient giant butcher", was certainly appropriately named. During the Pleistocene, it was the largest terrestrial predator in Australia.

On all other continents except Antarctica, there were large Pleistocene carnivorous mammals preying on large herbivorous mammals. However, apart from thylacines and marsupial lions there appears to have been very few large mammalian carnivores in Australia. **Megalania prisca** and its many carnivorous relatives in Australia, including the living Perenti and Gould's Goanna, may have been part of the reason for this.

No complete skeleton of **Megalania prisca** is known but from the partial ones that have been found, the maximum length was at least 7 m and the maximum weight about 600 kg. These animals were at least twice as long as their closest living relatives, the Komodo Dragon, a member of the genus **Varanus**. Species of **Varanus** are most diverse in Australia but also well represented in the modern faunas of Asia and Africa. This group of lizards is thought to be one of the most closely related to snakes. This is evident in several features such as the capacity for many of the bones of the skull to move with respect to one another, a common feature among snakes.

With its scimitar-shaped teeth, **Megalania prisca** was quite a capable carnivore. Perhaps it was even capable of killing the Pleistocene **Diprotodon optatum,** the largest potential prey species available to it. Like the living Komodo Dragon which can bring down an animal as large as a bullock by hamstringing it, **Megalania prisca** might have ambushed and killed **Diprotodon optatum.** But whether or not it could kill an animal this size, it certainly would have fed on dead individuals found as carrion. This is normal feeding behaviour for most of its living relatives.

Any statement about the geographic range of **Megalania prisca** within Australia must be tentative because of the scarcity of their remains. However, the few remains that are known suggest that it occurred in inland areas rather than the southern corner of the continent.

All records of **Megalania prisca** are confined to the Pleistocene epoch. It is not clear whether it died out near the end of the Pleistocene when there was widespread extinction of large marsupials and birds or whether they became extinct somewhat earlier.

CLASS: Reptilia
SUBCLASS: Lepidosauria
ORDER: Squamata
FAMILY: Varanidae
GENUS AND SPECIES: **Megalania prisca**

Compared with modern goanna

A giant python from southern Australia
Wonambi naracoortensis

Meredith J. Smith

Although snakes are found in many fossil deposits in North America, South America and Europe, the known fossil snakes throw little light on the evolution and radiation of snakes as a whole because most of the extinct forms are closely related to living species. One group without living representatives is the subfamily Madstoiinae of the family Boidae. Living boids, some of which are the largest living snakes, include the boas and pythons, nonpoisonous snakes that kill by constriction. The madstoiines also were large snakes. The subfamily is represented by specimens of Paleocene and Eocene age from Patagonia, late Cretaceous age from Madagascar and Eocene age from Egypt and Libya. Most recently, a fossil boid, **Wonambi naracoortensis**, was described from Pleistocene deposits in Australia and tentatively assigned to the Madstoiinae.

Most specimens of **Wonambi naracoortensis** have been found in Victoria Cave at Naracoorte, South Australia, but even here the species is rare. Only fifteen vertebrae, four broken ribs and two fragments of jaw have been found. Two vertebrae from each of two caves in southwestern Australia also appear to be referable to **Wonambi naracoortensis.**

Because the snake skull is fragile and seldom recovered from fossil deposits, vertebrae are more often used for identification of species. Snake vertebrae can be recognised by having on each vertebra three separate types of contact between successive vertebrae in the backbone: ball and socket joint; a mortise and tenon articulation; and lateral, paired, elliptical surfaces at the back of each vertebra which fit over matching surfaces at the front of the following vertebra. The vertebrae of **Wonambi naracoortensis** are large and seem cumbersome because of the very large surfaces for the hingeing of the ribs. In living boids the outermost parts of each vertebra are the lateral surfaces for contact between vertebrae whereas in the Madstoiinae the rib joint surfaces are the outermost. Another difference is that in living boids the lateral contact surfaces are nearly horizontal but in the Madstoiinae they are upturned, making an angle of about 25° from the horizontal. **W. naracoortensis** vertebrae resemble those of other madstoiines in overall shape and, more importantly, in such fine detail as the positions of tiny holes through which blood vessels entered the bone.

If **Wonambi naracoortensis** is, as its appearance suggests, closely related to other species of the Madstoiinae, then it is most likely that it colonized Australia before Australia drifted from Antarctica. Being found in Pleistocene deposits, **W. naracoortensis** survived much longer than other madstoiines, which were extinct before the Miocene.

Wonambi naracoortensis was about 5 m long and relatively thick-bodied. It probably was partly aquatic. Snakes of such large size are now found only in or near the tropics where the warm air and warm ground allow them to maintain a body temperature high enough for activity. The fauna found with **W. naracoortensis** in Victoria Cave suggests that the Pleistocene climate was not markedly different from the present climate. Hence the Pleistocene winters were probably cool and this boid must have been inactive for much of the year. Its growth rate would have been much slower than in living pythons. Reproduction, whether by laying eggs or giving birth to live young, would have been hazardous because embryonic development must have been possible for only a short period of the year. While further specimens of **W. naracoortensis** may clarify our understanding of its relationships, it is inevitable that its biology will remain speculative.

CLASS: Reptilia
ORDER: Squamata
SUBORDER: Ophidia
FAMILY: Boidae
GENUS AND SPECIES:
Wonambi naracoortensis Smith, 1976.

The world's largest bird
Dromornis stirtoni

Pat. V. Rich

Many Aboriginal legends allude to giant animals that no longer exist in Australia. Amongst these are several about "giant emus"; the Tjapwurong people of western Victoria, indeed, had a name for such feathered leviathans, 'mihirung paringmal', and oral tradition suggested that such birds were alive when volcanoes were still erupting in the Western District, a minimum of six to seven thousand years ago.

Such evidence is tantalizing, but does not prove, that such elephantine birds actually existed in the past. However, the fossilized bones of a whole series of giant birds discovered in Australia over the past century and a half, are conclusive evidence for the reality of a group of giant birds (Family: Dromornithidae) unique to this continent. They range in age from at least about 15 million years ago until 26,000 years ago, perhaps even much younger. Emphasizing that they certainly are **not** giant emus, these dromornithid birds have come to be called the Mihirung birds.

Some Mihirungs were only slightly larger than the living Emu, some gracile, some more ponderous. One form, **Dromornis stirtoni,** was truly of fantastic proportions, reaching perhaps 3 m or more in height and probably weighing more than 500 kg, rivalling or surpassing the elephant bird **(Aepyornis)** of Madagascar in size, and thus being the most weighty bird ever known.

Dromornis stirtoni is known from one locale near Alice Springs in the Northern Territory in rocks of Miocene age and lived alongside two other decidely smaller forms. Pollen collected from sediments of similar age in the area suggests that the climate was much wetter when **Dromornis** was alive, with forests of Southern Beech gracing the landscape. Evidently there were fresh-water lakes frequented by flamingoes, crocodiles, and lungfish as well — a very different scene from that of today.

Dromornis and the other Mihirungs are a unique group of birds with very deep lower jaws, "hoof"-like toe bones and a distinctly shaped bone connecting the lower and upper jaws, called the quadrate. Their skeletons suggest that they were plant eaters rather than carnivores because they lack recurved claws and a hook on the beak. But what they ate is a real mystery. Their jaws are capable of withstanding large forces and are very unlike the gracile jaws of Emus and Cassowaries.

Mihirungs may have been most successful in forest-environments that characterized much of Australia for much of the Cainozoic. As the forests contracted, grasslands expanded, and groups such as kangaroos diversified in these grasslands, the Mihirungs became extinct. What dealt the final blow to this once successful group is yet to be discovered.

CLASS: Aves
ORDER: Unknown
FAMILY: Dromornithidae
GENUS AND SPECIES:
Dromornis stirtoni Rich, 1979

Compared with modern Emu

Australia's oldest known monotreme
Obdurodon insignis

Michael Archer

Platypuses, like echidnas, lay eggs and have much about their structure that is more reptile-like that than of any other kind of living mammal. Together, the platypuses and echidnas comprise the primitive group of mammals called monotremes. And yet, until 1971, almost nothing of any significance was known about the fossil record of this obviously ancient group. In that year, a fossil tooth was found in 15 million year old fresh-water sediments from Frome Downs Station, South Australia. Although the tooth was broken and worn, it was clearly not like the tooth of any known kind of placental or marsupial mammal. For the rest of that year and most of 1972, the significance of the tooth was unclear.

In October, 1972, a second similar fossil tooth glistened in the bottom of a sieve. This one had come from 15 million year old fresh-water sediments in the Simpson Desert. Study of the two obviously very similar teeth made it clear that they showed a distinct resemblance to the juvenile teeth of the modern Platypus (**Ornithorhynchus anatinus**). These teeth in the living animal are poorly known because they have only been found in a few of the rarely collected juveniles. Normally, by the time the juveniles reach adulthood, they have shed their vestigial teeth using instead horny pads to crush their food.

Despite the similarity between the fossil teeth and the vestigial teeth of the modern platypus, there are also many differences. The tooth crown of the fossil is higher and somewhat differently shaped. Its larger roots suggest that it was of considerably greater value to this extinct platypus than were the teeth of its living relative. As a result, in 1975, this most venerable of monotremes was given its own distinctive generic name **Obdurodon,** which means in Latin "enduring tooth".

Later discoveries in 1977 (at the Simpson Desert site) of fragments of a pelvis and a jaw, both referable to **Obdurodon insignis,** enabled partial reconstruction of this playtpus. It appears to have had a much smaller bill and probably was less efficient as an excavator of burrows than the living species. It is also possible to speculate about its habits. During the 1977 effort to recover more of this fascinating beast, many tonnes of fossil-rich sediments were processed. As a result, although only two fragments referable to the fossil platypus were recovered, dozens of koalas, ringtail possums and other decidely **non** aquatic mammals were also found. If this extinct platypus was habitually aquatic like its living relative, wouldn't it seem likely that it would have been more common than koalas as a fossil in these fresh-water deposits?

Assuming that it was not, therefore, as habitually aquatic as its living relative, the artist has placed our short-billed friend on shore, perhaps as a predator of frogs or crustaceans along the margins of the fresh-water lake. Hopefully future discoveries will enable us to refine our understanding about this intriguing monotreme.

CLASS: Mammalia
SUBCLASS: Monotremata
FAMILY: Ornithorhynchidae
GENUS AND SPECIES:
Obdurodon insignis, Woodburne and Tedford, 1975.

The giant echidnas
Zaglossus species

Neville Pledge

The New Guinea Highlands are the home of the only surviving species of **Zaglossus**, the Mountain Echidna **(Z. bruijni)**. However, fossil species of **Zaglossus** have been found in most Australian states. At least three extinct species have been recognised: **Z. hacketti, Z. robustus,** and **Z. ramsayi.** (The last is the species illustrated.) Fossils of the modern **Z. bruijni** are also known.

Zaglossus robustus, from the Canadian Deep Lead mine at Gulgong, New South Wales, is the oldest of these species. It lived about three million years ago (in the late Pliocene). The others are generally of late Pleistocene age. The Tasmanian specimens are only about 20,000 years old.

The modern species of **Zaglossus** is larger than the Common Echidna **(Tachyglossus aculeatus).** The fossil species of **Zaglossus** were even bigger. **Zaglossus hacketti,** known from only a few bones from Mammoth Cave, Western Australia, was possibly the largest monotreme that ever lived. **Zaglossus** species are distinguished from Common Echidnas by their more erect stance and long, down-curved snout. In the modern species, the spines are also less obvious than they are in the Common Echidna, being shorter, fewer, and partly hidden in the fur.

Zaglossus ramsayi is the best-known fossil species, being represented by many isolated bones and a partial skeleton with an almost perfect skull. The skull is smooth, rounded and very fragile, much like that of a bird in general appearance. Its most notable feature is its long, sturdy and gently curved snout, which differs from the snout of the living **Z. bruijni** in its sturdiness, relative shortness, greater breadth, and straightness. This suggests that **Z. ramsayi** differed in diet and way of life from the living species.

Unlike the Common Echidna which lives mostly on ants and termites, **Z. bruijni** is reported to eat mainly worms. These it forages from the leaf litter and humus of the forest floor. The shape of their palates suggests that **Z. ramsayi** and **Z. robustus** were able to eat fairly large food items and the robustness of the bones of their forelimbs and snouts indicates great strength for digging and probing amongst rocks and logs. Perhaps these giant echidnas ate grubs and beetles and other invertebrates as well as ants, termites and worms.

Species of **Zaglossus** appear to have become extinct on mainland Australia by the end of the Pleistocene. Perhaps this was the result of natural ecological changes and pressures which selected for the smaller surviving echidna, **Tachyglossus aculeatus.** Probably we will never know what happened.

CLASS: Mammalia
SUBCLASS: Monotremata
FAMILY: Tachyglossidae
GENUS AND SPECIES: **Zaglossus ramsayi.**

Compared with modern Echidna

Australia's largest known carnivorous marsupial
Thylacinus potens

Michael Archer

When Europeans arrived in Australia, the Tasmanian Thylacine was Australia's largest marsupial carnivore. So, of course, with guns, baits, snares and fear of loss of money, they exterminated it. It was one of many species we pushed over the brink. But this particular act of barbarism was not like all the others. It was much worse. With the Tasmanian Thylacines's extermination, a whole family of mammals was eradicated. That extinction seems all the more tragic with the realisation that this magnificent species, **Thylacinus cynocephalus,** had survived for almost 4 million years — until our arrival.

Thylacinus potens, the "powerful" thylacine, is the largest and only other species of the family known. It appears to have preceded the Tasmanian Thylacine by about 4 to 6 million years. It is only known from one 8-10 million years old fossil deposit near Alice Springs from which have also come crocodiles, flamingoes, ringtail possums, many large herbivorous marsupials and **Dromornis stirtoni,** the world's largest known bird.

This thylacine appears to have had a somewhat shorter, broader head than the modern Tasmanian species and was presumably slightly more heavily built than its modern counterpart. Its teeth, although not quite as specialized for meat-cutting as those of **Thylacinus cynocephalus,** were certainly those of a carnivore. Running down and killing prey would have been no problem for this pouched animal.

As yet the fossil record of thylacines is too poor to enable us to determine how closely related they are to the other groups of marsupials. However, studies of the properties of blood proteins, taken from museum skins of Tasmanian Thylacines, suggest they are most closely related to dasyurid marsupials.

The oldest known thylacine is represented by a tooth from a 14-15 million year old deposit in northwestern Queensland. This suggests that the family will prove to have had a very long history — a history that regrettably ended with our arrival.

CLASS: Mammalia
INFRACLASS: Marsupialia
FAMILY: Thylacinidae
GENUS AND SPECIES: **Thylacinus potens.**

Relic of a lost tooth mine
Glaucodon ballaratensis

Michael Archer

In 1914, a Mr. J. Marshall presented to the National Museum of Victoria a very beautifully preserved fossil dentary that still retained two of its original ten teeth. He said that it had been recovered from a depth of "50 feet" in a well 2.5 km. northwest of Smeaton, near Ballarat, Victoria. If the significance of the dentary had been realized at the time it was turned over to the Museum, it is just possible that **Glaucodon ballaratensis** would now be represented by more than this single dentary with its two teeth.

As things turned out, perhaps in part because no interest in the fossil was expressed at the time, the well was filled in with rocks and its precise location lost. The sad result is that we will probably never see any more of this fascinating animal, nor of any possible contemporaries that may have been buried with it in the same sediments.

On the basis of the dentary and the two teeth which were preserved, **Glaucodon ballaratensis** was clearly a fascinating and highly specialized carnivorous dasyurid. Most of its features resemble those of the living Tiger Quolls **(Dasyurus maculatus)** but some resemble those shown by the living Tasmanian Devil **(Sarcophilus harrisii)**. Accordingly, since its scientific description in 1857, it has generally been regarded as more or less intermediate in structure between the Quolls and the Devils. Its age is also in doubt because the sediments from which it came cannot now be examined. However, it is likely to be late Pliocene to early Pleistocene.

In terms of its evolutionary significance, there seem to be two equally probable possibilities. Either it was an animal related to the ancestors of the Tasmanian Devil or it was a specialized highly carnivorous type of Quoll that was developing Devil-like features. Unfortunately, the decisive information now lies buried somewhere beneath 17 metres of rock.

CLASS: Mammalia
INFRACLASS: Marsupialia
FAMILY: Dasyuridae
GENUS AND SPECIES:
Glaucodon ballaratensis Stirton, 1975.

The marsupial lion
Thylacoleo carnifex

Eileen Finch

During the Pleistocene, the large **Thylacoleo carnifex** roamed most of Australia except the arid centre, in company with large marsupials such as the giant wombat (**Phascolonus gigas),** giant kangaroos (e.g. species of **Procoptodon**) and **Diprotodon australis.**

This was a most unusual animal. It was first known from a long ridge-like tooth found in the 1830s in the Wellington Caves, New South Wales, the discovery at first utterly confounded England's famous palaeontologist of the day, Richard Owen. Later discovery of skull fragments suggested affinities with the possums, but the unique features of this animal were very un-possum-like. Of the three pairs of upper incisors, the first was very large, the two lower incisors were forwardly inclined, and, with the enlarged upper incisors, formed a powerful set of pincers. All other teeth were small except the enormous (4-5 cm) carnassial or cutting premolar tooth in each jaw. Only one upper and two lower molars remained behind the premolars, the others having been reduced to the point of loss.

What could have been the feeding habit of this huge "possum"? Owen, indicating his eventual conclusion, had named it **Thylacoleo carnifex** (flesh-eating marsupial lion) and described it as "one of the fellest . . . of predatory beasts." Others, stressing the herbivority of most possums and the Marsupial Lion's lack of large canines for killing prey, concluded that **Thylacoleo** was a plant feeder that used its enormous premolars to cut stems and tough fruits. Owen countered with the fact that herbivores need grinding molars which **carnifex** lacked. Other ideas were proposed including that the animal was a bone-crushing scavenger. However, modern experiments with bones and teeth have not provided much support for this idea. The most recent reconstructions of jaw musculature, along with work on jaw mechanics and studies of the microscopic wear patterns on the carnassial teeth have revealed that the powerful jaws of **Thylacoleo carnifex** were, in fact, as Owen surmised, well-adapted to kill. Its incisors may have been used like knives to kill, and its carnassials to rend the dead prey of its flesh or to dismember the carcass into bite-size bits.

In the 1950s the first postcranial skeleton material was unearthed in South Australia. Since then, more has been found in New South Wales. The vertebral column was strong, yet flexible, while the limbs were long and powerful. All digits were clawed but the "thumb" supported a huge compressed claw 3-4 cm in length. The forelimb was obviously an efficient striking and holding weapon, the hindlimb was possum-like with an opposable first toe. This sort of foot serves to provide grip and balance in arboreal possums like the Common Brushtail Possum (**Trichosurus vulpecula).**

Certainly its limbs were well adapted for medium-paced running and prey-catching, but to what extent this Leopard-sized carnivore used its possum-like hind foot to climb is uncertain. The answer must await further study of the anatomy of each of the bones of the whole skeleton.

CLASS: Mammalia
INFRACLASS: Marsupialia
ORDER: Diprotodonta
FAMILY: Thylacoleonidae
GENUS AND SPECIES: **Thylacoleo carnifex.**

Killing Procoptodon

A unique trunked giant
Palorchestes azael

Tim Flannery

Perhaps no animal is as suited to have inspired the legend of the bunyip as **Palorchestes azael**. Although a herbivore, it must have been a fearsome sight. The largest of the three known species of the genus, **Palorchestes azael,** was the size of a bull. Its exceptionally massive forearms were equipped with razor sharp, rapier-like claws up to 12 cm long, and its bizarre head was crowned with a longish trunk. Surely this sight would have been enough to give any newly arrived Aboriginals second thoughts about settling. And there is little doubt that **Palorchestes azael** and humans were acquainted. In fact, they co-existed in Australia between about 40-20,000 years ago.

The species of **Palorchestes** have a long history. The oldest remains from near Riversleigh Station in northwestern Queensland are about 14 million years old. Remains about 6 to 8 million years old are known from a site near Alice Springs. However, remains of species of **Palorchestes** less than 5 million years old are restricted to eastern Australia. The species of **Palorchestes** may have been unable to survive in central Australia because about 10 to 5 million years ago this region began to dry out. With the drying and spread of grassland the forests dwindled.

The species **Palorchestes** are related to the diprotodontids, a group that includes **Diprotodon optatum,** the largest of the marsupials. However, they are so different from the diprotodontids that recent workers tend to place them in their own family, the Palorchestidae.

We can only speculate about the habits of **Palorchestes azael.** The rarity of its remains suggests that it was a solitary animal. Its powerful clawed limbs were not adapted to digging but may have been used to strip branches from trees or uproot bushes in search of tubers and roots. Its high crowned teeth suggest that it ate abrasive food such as bark or roots. Unfortunately we must await a detailed study of the skeletal anatomy of this fascinating animal before more can be determined about its lifestyle.

CLASS: Mammalia
INFRACLASS: Marsupialia
ORDER: Diprotodonta
FAMILY: Palorchestidae
GENUS AND SPECIES: **Palorchestes azael.**

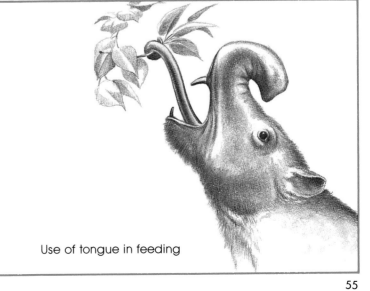

Use of tongue in feeding

An ancient Territorian
The Bullock Creek Neohelos

Michael D. Plane

Species of **Neohelos** (which means "new bump" in reference to a distinctive cusp on the premolar) are medium-sized members of the sub-family Zygomaturinae of the family Diprotodontidae. This family of browsing herbivores included small wombat-sized to rhinoceros-sized forms which roamed all areas of the Australian continent from Tasmania to Papua New Guinea for at least the last 23 million years.

The first known species of the genus, described by the late Professor R. A. Stirton, came from the Kutjumarpu fauna of the Lake Eyre Basin, and is known only from isolated teeth. Before Stirton's description of this species was published in 1967, but unfortunately not before his untimely death, a happy accident brought to light new and significant fossil sites at Bullock Creek in the Northern Territory. A palaeontologist interested in Cambrian trilobites was searching in limestone rocks south of Montijinni, a famous trilobite locality, when he spotted what appeared to be a well-preserved mammal tooth. History does not record his thoughts of the moment, but they may well have been: "I've either discovered the world's oldest mammal, the first trilobite tooth or a new Tertiary mammal site!"

In fact, he had just discovered an important new Tertiary mammal site. That tooth was the first of many discoveries from the hard white Miocene limestone of the Bullock Creek sites. Following slow and painstaking preparation by a technician, many beautiful, black and marvellously preserved skulls, jaw and limbs emerged. And of these, the vast majority represented a second, new species of **Neohelos.**

But the discovery of new species of **Neohelos** didn't stop in the Territory. The freshwater Carl Creek limestone of Riversleigh Station, along the banks of the Gregory River, were also known to produce fossils of aquatic invertebrate and vertebrate animals. Riversleigh mammal fossils, however, are rare. So it was a very exciting moment when a partially exposed upper premolar and first molar of yet a **third** new species of **Neohelos** were spotted in a block of Riversleigh limestone.

Together, the material from all three sites provides us with a tantalising glimpse of these quadrupedal herbivores. Rather like large long-legged wombats, they probably browsed on damp vegetation in the forests surrounding lakes and river flats which covered much of Australia in middle and late Miocene times.

At night, in the trees above them, possums of many kinds would have scampered in the branches feeding while keeping a wary lookout for the carnivorous marsupial lions (species of **Wakaleo,** small early cousins of the Pleistocene **Thylacoleo carnifex**). Still higher overhead, bats (such as an ancestral ghostbat) would have hunted silently between the trees. At dusk, small, primitive kangaroos (such as **Wabularoo naughtoni**) would have come to drink at the water's edge. And in the water, poised in wait for the intrusion of the muzzle of a thirsty mammal, lay crocodiles. Fortunately for the palaeontologist, many of these muzzles must have belonged to the thirsty but incautious species of **Neohelos.**

CLASS: Mammalia
INFRACLASS: Marsupialia
FAMILY: Diprotodontidae
GENUS: Neohelos,
            species not yet determined.

A rhinoceran relative of the wombats
Zygomaturus trilobus

Ken Aplin

Coexisting with **Diprotodon australis** during the Pleistocene period was a second, equally bizarre member of the diprotodontid family. This was **Zygomaturus trilobus,** a bullock-sized herbivore that ranged widely throughout the better-watered areas of southern and eastern Australia.

Much of our knowledge concerning the anatomy of this species comes from a number of beautifully preserved skeletons discovered in 1912 at Mowbray Swamp, in northwestern Tasmania. From this material we are able to reconstruct **Zygomaturus trilobus** as a massive, solidly built beast, standing 1.5 m. tall at the shoulder. It was without question a ground dwelling animal, moving on all fours with a slow but deliberate gait. The fore and hind-limbs were heavily muscled but nonetheless quite elongate. The fore-feet of **Zygomaturus trilobus** were exceedingly broad, with the five fingers ending in elongate but flattened claws. By comparison, the hind-feet were quite slender, and, except for the outermost digit, the toes were weakly clawed. As in the modern wombats, the hind-foot was rotated inwards such that the animal walked on the side of its foot. In common with other diprotontids, **Zygomaturus trilobus** had only a short, strap-like tail.

The skull of **Zygomaturus trilobus** is both absolutely and proportionately massive. However, as is often the case with such large animals, much of this volume was taken up by an extensive network of air-filled sinuses. These would have served to lessen the otherwise crippling weight of such a massively boned and muscled head. Despite the large size of its head, the brain of **Zygomaturus trilobus** was evidently quite small. **Zygomaturus trilobus** was, in all likelihood, a gentle giant by nature. Opinion varies as to the anatomy of the snout of **Zygomaturus trilobus**. To some the presence of paired, rugosities on its upper surface indicate the development of keratinous horns akin to those of the Rhinoceros. To others, the overall construction of the bony rostrum suggests the presence in life of a short, but highly mobile trunk. The former of these alternatives is indicated in the illustration.

Zygomaturus trilobus probably existed in southern Australia throughout the Pleistocene period. Although several, slightly older species of **Zygomaturus** have been described, it is uncertain which, if any, of these was the progenitor of the Pleistocene form. Further back in time, the diprotodontid family as a whole probably shared a common ancestor with the wombats.

Several fossil localities containing the remains of **Zygomaturus trilobus** date to as recent as 19,000 years ago. Although it is therefore quite likely that the early Aboriginals were well acquainted with this magnificent beast, the nature and extent of any interaction between the two parties is unknown. Indeed, the possible role of the Aboriginal population in the eventual extinction of both **Zygomaturus trilobus** and its gargantuan relatives remains one of the most hotly debated topics in Australian palaeontology.

CLASS: Mammalia
INFRACLASS: Marsupialia
ORDER: Diprotodonta
FAMILY: Diprotodontidae
GENUS AND SPECIES:
Zygomaturus trilobus Owen, 1859.

Cheeky Giant of the Pliocene
Euryzygoma dunense

Ken Aplin

In 1912 a noteworthy discovery was made during the sinking of a well on a property near Brigalow in the Darling Downs region of southeastern Queensland. At a depth of about 20 m, the well-sinkers found an almost intact fossilized skull of a gigantic and highly unusual extinct marsupial, the diprotodontid **Euryzygoma dunense**.

The discovery of fossil bones in the Brigalow area was by no means an exceptional event (the Darling Downs were well-known as an area rich in vertebrate fossils). However, this particular skull was of outstanding interest both for its remarkable completeness, and for its truly extraordinary proportions. Sprouting from each side of the head, at about the level of the eye-sockets, are paired bony excrescences of remarkable breadth and robustness. These give the skull a decidedly monstrous, but at the same time, somewhat ludicrous appearance.

Although the Brigalow specimen provided the first information on the skull of **Euryzygoma dunense,** the species itself had been known since the late 1880s. However, working from jaws and teeth alone, its original describer, Mr. C. W. De Vis, had considered the species to belong within the relatively unspecialized diprotontid genus **Nototherium**. Only with the discovery of the Brigalow specimen did the full peculiarity of De Vis' species become apparent. Accordingly, Mr. H. A. Longman, De Vis' successor at the Queensland Museum, established of new genus, **Euryzygoma,** for the fossil species.

Among placental mammals, the widely flaring cheekbones of **Euryzygoma dunense** are paralleled only among certain North American rodents of the pocket-gopher family. In these animals, the large cheekbones support large, fleshy pouches which are used as temporary food-stores. Longman considered that similar cheekpouches may have been present in **Euryzygoma dunense**. Presumably, with its capacity to store food in its cheekpouches, the animal would have had to spend less time feeding at or near the crocodile infested, prehistoric waterways of the eastern Australia.

Other possible interpretations of the flaring cheekbones include their potential role in sexual or territorial displays (some skulls of **Euryzygoma dunense** appear to lack the flaring cheekbones, suggesting the possibility of sexual dimorphism), or as a defense mechanism (to ram or slice at ravenous marsupial lions?). Another, and perhaps even more fanciful suggestion, would have **Euryzygoma dunense** as a marsupial counterpart of the Howler Monkey, its cheeks and throat supporting huge, inflatable resonating chambers.

Although **Euryzygoma dunense** is best known for its occurrence in the deposits of the Darling Downs, its remains have also been found at a number of other sites in northern Queensland and northwestern New South Wales. Together, these sites indicate a time range for the species from the early Pliocene to the mid-Pleistocene (approximately 4.5 to 1.0 million years ago). To date its remains have not been discovered in any site dated by the radiocarbon method. **Euryzygoma dunense** would thus appear to have become extinct well prior to 40,000 years ago and the coming of Aborigines to Australia.

CLASS: Mammalia
INFRACLASS: Marsupialia
ORDER: Diprotodonta
FAMILY: Diprotodontidae
GENUS AND SPECIES:
Euryzygoma dunense De Vis, 1887.

The largest marsupial
Diprotodon optatum

Thomas H. V. Rich

Diprotodon optatum, which means the "southern" animal with "two forward-projecting teeth", was the first fossil mammal named from Australia. When initially recognised in 1838, only a few teeth and a bit of jaw were known. It was not until 1893 that feet were finally found. And despite the fact that remains of **Diprotodon optatum** are among the most common of extinct marsupials found in Australia many details of the anatomy of the skull are still uncertain.

Nearly 3 m long and 2 m high at the shoulder, **Diprotodon optatum** was the largest marsupial that ever lived. It was also one of the last members of its family of browsing, four-footed marsupial herbivores, the Diprotodontidae. This family flourished in the late Tertiary but declined in diversity during the Pleistocene when **Diprotodon optatum** lived. It was during this decline that the kangaroos increased dramatically in number of species and average size. Perhaps the rise of these two-footed herbivores was instrumental in the decline of the diprotodontids.

Although the family as a whole may have been in decline during the Pleistocene, the onset of arid conditions at the end of the Cainozoic may have been a factor that brought about the appearance of **Diprotodon optatum.** Remains of this species have been found all over Australia except Tasmania and the western part of Western Australia. **Zygomaturus trilobus,** another diprotodontid, occurs in both of those areas and along the coastal region of southeastern Australia. Although their ranges overlap, this general difference in their distribution within Australia suggests that **Diprotodon optatum** have been adapted to more open habitats while **Zygomaturus trilobus** favoured more forested ones. Because species of **Zygomaturus** are known from the late Miocene to the late Pleistocene and species of **Diprotodon** are unknown prior to the Pleistocene, it is tempting to relate the appearance of species of **Diprotodon** to the spread of less wooded grassland habitats at the end of the Cainozoic.

Diprotodon optatum was a member of a group of large marsupials and birds that existed during the Pleistocene in Australia and became extinct between 15,000 and 25,000 years ago. The cause of demise of these many different species has been attributed to changes in climate or to the action of Aboriginals. If Aboriginals were responsible, they may have brought about the extinction directly by hunting or indirectly by altering the landscape through such practices as the deliberate burning of large areas of land.

CLASS: Mammalia
INFRACLASS: Marsupialia
ORDER: Diprotodonta
FAMILY: Diprotodontidae
GENUS AND SPECIES: **Diprotodon optatum**

Alternate reconstruction with snout

Largest of the wombats
Phascolonus gigas

Lyn Dawson

Phascolonus gigas was the largest, and most unusual, of several large extinct wombats that inhabited Australia during the Pleistocene. **P. gigas** was fundamentally similar to modern wombats, but it was not closely related to any particular kind of living wombat. It probably evolved, in the early Pliocene, from a **Vombatus**-like ancestor which is, as yet, unknown.

A nearly complete skeleton of **Phascolonus gigas** is known from Lake Callabonna, South Australia. Although its skull is approximately 40 cm long (about the size of a cow's skull), the rest of the skeleton indicates that the animal stood only about 1 m tall, with very short and robust limbs.

Phascolonus gigas had most unusual teeth. The molar teeth are bilobed, curved and open-rooted, like the molars of all other Pleistocene and living wombats. But the upper incisors are very long, curved and extremely wide, superficially resembling those of **Diprotodon.** These incisors are so improbable in a wombat skull that even as recently as 20 years ago it was suggested that they had been placed in the Callabonna skull as a hoax. This has now been definitely refuted.

The feet of **Phascolonus gigas**, known from a few digits, were apparently not as specialised for burrowing as those of modern wombats, being also more like those of **Diprotodon.** If it did burrow, **P. gigas** would have been the largest fossorial animal ever known.

Teeth and jaw fragments of **Phascolonus gigas** are known from Pleistocene fossil deposits from Queensland, New South Wales, Victoria and South Australia. A few teeth fragments from deposits near Allingham in Queensland and Merriwa in New South Wales indicate that a species of **Phascolonus** occurred in the early Pliocene.

During the Pleistocene, **Phascolonus gigas** was apparently most abundant around inland lakes and watercourses, such as Cooper's Creek, Lake Eyre and Lake Callabonna. This large wombat may have foraged along the banks of lakes and streams, possibly using its large adze-like incisors to dig in mud for roots or tear bark from trees and shrubs to supplement a basic diet of grass and leaves.

CLASS: Mammalia
INFRACLASS: Marsupialia
FAMILY: Vombatidae
GENUS AND SPECIES:
Phascolonus gigas Owen, 1858.

Compared with modern wombat

A different kind of kangaroo
Wabularoo naughtoni

Tim Flannery

Wabula means "long time ago" in the language of the Aboriginals of Queensland's gulf country, and **Wabularoo** is therefore a particularly appropriate name for one of Australia's oldest fossil kangaroos. Fossils of this species have been found preserved in limestone on Riversleigh Station, northwestern Queensland, and are probably about 14 million years old. At this time, most of Australia received a higher rainfall and the area around Riversleigh may well have supported much wetter forests than it does today.

Along with the remains of **Wabularoo naughtoni,** the fossil bones of many other extinct kinds of animals, including five other species of kangaroos, have been found in the Riversleigh deposit. All of the kangaroos from this remote period of Australia's history were small primitive types. **Wabularoo naughtoni,** although probably weighing only 5-10 kg, was one of the larger ones. Among the other kinds of kangaroos found at Riversleigh are rat-kangaroos similar to the living bettongs and some very primitive true kangaroos belonging to an extict subfamily. The smallest of the Riversleigh kangaroos, a relative of **Wabularoo naughtoni,** was only the size of a bandicoot.

Wabularoo naughtoni appears to be related to the rat kangaroos (potoroids), but is sufficiently different to be placed in a separate subfamily. One major difference is the structure of its molars which resemble those of the true kangaroos. It is probable that **Wabularoo naughtoni** represents an early rat-kangaroo "experiment" to occupy the ecological niche of today's browsing kangaroos such as the forest wallabies of New Guinea. **Wabularoo naughtoni** and its relatives became extinct when the more advanced browsing kangaroos evolved in the late Miocene period, 5-10 million years ago.

CLASS: Mammalia
INFRACLASS: Marsupialia
ORDER: Diprotodonta
FAMILY: Potoroidae
GENUS AND SPECIES: **Wabularoo naughtoni.**

A giant Australian tree kangaroo
Bohra paulae

Tim Flannery

Most people think of kangaroos as inhabitants of Australia's grassy plains and woodlands. Few are familiar with the very different kinds that live in the canopy of the tropical rainforests of north Queensland and New Guinea: the tree kangaroos. There are seven species of living tree kangaroos all belonging within the genus **Dendrolagus.** They are medium sized kangaroos, the largest weighing about 18 kg.

Until recently, the tree kangaroos were thought to have always been restricted to the tropics. Then in 1982, fossilised limb bones of a giant kind of tree kangaroo were recognised amongst bones in the Australian Museum's collection from the Wellington Caves, New South Wales. Australian fossil marsupials were first discovered in the Wellington Caves, and described as early as 1831. But the deposits are so rich that despite this early start, they have continued to yield new species up to the present day. **Bohra paulae**, as this fossil tree kangaroo was named, was a large animal, weighing perhaps as much as 60 kg. Although this extinct species was more primitive than living tree kangaroos, its bones show many of the adaptations possessed by living tree kangaroos that enable them to climb into the treetops.

The bones of **Bohra paulae** are Pleistocene in age, somewhere between 2 million and 10,000 years old. Unfortunately we don't know when **Bohra paulae** became extinct. It may have died out in response to the increasing aridity of Australia and the subsequent changes in the forests which occurred during the Pleistocene period. It is even possible that **Bohra paulae** survived until the coming of the Aboriginals. They would have found 60 kg of delicious flesh perched helplessly in a tree too good a meal to resist.

We can only guess at the nature of the forest in which **Bohra paulae** lived. Today, tree kangaroos live only in rainforest. Because the bones of very few (if any) rainforest inhabiting species have been found in the Wellington Caves deposits, it is possible that **Bohra paulae** inhabited drier types of forest.

Modern tree kangaroos are browsers, specialising in eating the leaves of trees. Unfortunately, the teeth of **Bohra paulae** are unknown, so we do not know what sort of food it ate. Hopefully the continuing work at Wellington Caves will result in the discovery of further remains of **Bohra paulae** and in so doing enable more to be determined about the habits and habitat of this largest and most enigmatic of the tree kangaroos.

CLASS: Mammalia
INFRACLASS: Marsupialia
ORDER: Diprotodonta
FAMILY: Macropodidae
GENUS AND SPECIES: **Bohra paulae**

Compared with modern tree kangaroo

A gigantic relative of the banded hare wallaby

Troposodon kenti

Tim Flannery

The nearest relative of species of **Troposodon** is the living Banded Hare Wallaby (**Lagostrophus fasciatus**). This tiny animal, weighing 1-2 kg, inhabits dense scrub and survives today on only two islands off the coast of Western Australia. In contrast **Troposodon kenti** probably weighed in excess of 100 kg and its remains have been found in fossil deposits about 2 to 4 million years old associated with species which suggest that it inhabited savannah and woodland. Of the three localities where the remains of **Troposodon kenti** have been found, two are in eastern Australia and one is in northern South Australia.

The teeth of **Troposodon kenti** suggest that it was probably a browser, eating the leaves and twigs of bushes and herbs. After the Banded Hare Wallaby, the closest relatives of **Troposodon kenti** are the gigantic extinct short-faced kangaroos. Many of these were also browsers and began to increase in number and diversity about two million years ago, at the beginning of the Pleistocene period. It was at about this time that **Troposodon kenti** became extinct. Perhaps it was this rapid radiation of closely related species of similar habits which drove **Troposodon kenti** into extinction.

CLASS: Mammalia
INFRACLASS: Marsupialia
ORDER: Diprotodonta
FAMILY: Macropodidae
GENUS AND SPECIES: **Troposodon kenti**

Short-faced giant among kangaroos
Procoptodon pusio

Tim Flannery

From the time Europeans first encountered marsupials, they noted similarities between them and the more familiar placental mammals. Thus, we have our marsupial "cats", "tigers" and "bears". If one had to elect a "marsupial hominid", it would undoubtedly be one of the species of **Procoptodon**, for in many ways, these extinct giant kangaroos resemble hominids (the family containing ourselves, **Homo sapiens**). Both groups possess forward-pointing eyes, shortened faces, heavily enamelled molars, highly mobile arms, are bipedal and are of similar size.

Four species of **Procoptodon** are known, one of which, **Procoptodon goliah**, was the largest kangaroo ever. The hindlimbs of the species of **Procoptodon** are very unusual; they are elongate, perhaps suggesting that they could attain considerable speed. The foot possesses only a single large toe, ending in a hoof-like nail.

The nearest relatives of the species of **Procoptodon** are the species of **Simosthenurus**, another group of extinct short-faced kangaroos. However, rather than eating grasses, as the species of **Procoptodon** did, the species of **Simosthenurus** probably ate less abrasive food such as leaves and twigs of bushes and herbs. The nearest living relative of both of these groups of kangaroos is the diminutive Banded Hare Wallaby from Western Australia.

Procoptodon pusio is the smallest and least specialised species of **Procoptodon**, and was probably about the size of a living Grey Kangaroo. Its remains have been found in sediments of Pleistocene age (2 million — 10,000 years ago) in southeastern Queensland and eastern New South Wales. It probably inhabited woodland and savannah areas. It is not known when **Procoptodon pusio** died out. It may have died out before the more advanced species of **Procoptodon**, or may have become extinct with them with the coming of the Aboriginals. Whatever the case, it is a great pity that since the species of **Procoptodon** have become extinct, we cannot look into a "living evolutionary mirror" and see ourselves reflected in marsupial form.

INFRAORDER: Marsupialia
ORDER: Diprotodonta
SUPERFAMILY: Macropodoidea
FAMILY: Macropodidae
GENUS AND SPECIES: **Procoptodon pusio**

Fresh-water dolphins from central Australia

Ewan Fordyce

An unusual occurrence of extinct Australian animals is that of dolphins from the Lake Frome district of South Australia. Bones of about six individuals have been found in fresh-water sediments of middle Miocene age (about 14-16 million years old). The fossils comprise mostly hard, erosion-resistant earbones, but also skull fragments, teeth, vertebrae and limb-bones. Earbone structure indicates that the dolphins represent a species of the extinct family Rhabdosteidae (also known as Eurhinodelphidae), and belong with toothed whales, dolphins and porpoises (Odontoceti) in the mammalian Order Cetacea (which includes all whales, dolphins and porpoises). The Australian fossils are too incomplete to be able to identify the particular species. Hopefully, it will not be long before a fossil skull or other sufficiently distinctive portion of the skeleton is found.

Rhabdosteids were small to medium-sized dolphins with long beaks. Elsewhere, they are known from North and South America and Europe, from Lower and Middle Miocene rocks (about 15-25 million years old). The South Australian fossils provide the first record of rhabdosteids from Australasia, and represent their only record in fresh-water sediments. The number of specimens found in the Lake Frome district suggests that the occurrence is not an accidental one. These dolphins were probably important residents of the Australian middle Tertiary fresh-water ecosystems. They probably invaded central Australian fresh-waters from the sea via a large river like the Murray, although it is uncertain when this occurred.

The Australian rhabdosteids were probably similar in appearance and habits to four unrelated species of "river dolphins" found today in Asia and South America. On the basis of the size and structure of the fossils and their probable similarity to these modern "river dolphins", the extinct Australian form may have been 2 to 3 m long, with a light-coloured streamlined body, long beak, flexible neck, fairly long forelimbs, and a dorsal fin. They probably lived on fish which, judging from the abundance of fish bones in the same fossil deposits, must have been very common.

CLASS: Mammalia
INFRACLASS: Eutheria
ORDER: Cetacea
FAMILY: Rhabdosteidae
GENUS AND SPECIES: Unidentified.

References

Introduction

Archer, M. and Bartholomai, A., 1978. Tertiary mammals of Australia: a synoptic review. **Alcheringa 2:** 1-19.

Archer, M., 1981. A review of the origins and radiations of Australian mammals. Pp. 1437-88 in "Ecological biogeography in Australia" ed by A. Keast, W. Junk: The Hague.

Bartholomai, A., 1970. New lizard-like reptiles from the early Triassic of Queensland. **Alcheringa 3:** 225-34.

Campbell, K. S. W. and Bell, M. W., 1977. A primitive amphibian from the late Devonian of New South Wales. **Alcheringa 1:** 369-81.

Kemp, A., 1982. Australian Mesozoic and Cenozoic lungfish. Pp. 133-43 in "The fossil vertebrate record of Australasia" ed by P. V. Rich and E. M. Thompson. Monash University Offset Printing Unit: Melbourne.

Long, J., 1982. The history of fishes on the Australian continent. Pp. 54-85 in "The fossil vertebrate record of Australasia" ed by P. V. Rich and E. M. Thompson. Monash University Offset Printing Unit: Melbourne.

Molnar, R., 1983. Paleozoic and Mesozoic reptiles and amphibians from Australia. In "Vertebrate zoogeography and evolution in Australasia" ed by M. Archer and G. Clayton. Hesperion Press: Perth.

Rich, P. V., 1979. The Dromornithidae, a family of large, extinct ground birds endemic to Australia. **Bur. Min. Res. Bull. 184:** 1-196.

Rich, P. V., 1982. An all too brief and superficial history of Australian vertebrate palaeontology. Pp. 2-26 in "The fossil vertebrate record of Australasia" ed by P. V. Rich and E. M. Thompson. Monash University Offset Printing Unit: Melbourne.

Ricek, E. F., 1979. Lower Cretaceous fleas. **Nature 227:** 746-47.

Thulborn, R. A. and Wade, M., 1979. Dinosaur stampede in the Cretaceous of Queensland. **Lethaia 12:** 275-9.

Wade, M., 1979. Tracking dinosaurs: The Winton excavation. **Aust. Nat. Hist. 19:** 286-91.

Warren, A., 1972. Triassic amphibians and reptiles of Australia in relation to Gondwanaland. **Aust. Nat. Hist. 17:** 279-83.

Warren, J. W. and Wakefield, N. A., 1972. Trackways of tetrapod vertebrates from the Upper Devonian of Victoria, Australia. **Nature 238:** 469-70.

Warren, A., 1977. Jurassic labyrinthodont. **Nature 265:** 436-37.

Warren, A., 1982. Australian fossil amphibians. Pp. 146-57 in "The fossil vertebrate record of Australasia" ed by P. V. Rich and E. M. Thompson. Monash University Offset Printing Unit: Melbourne.

Rolfosteus canningensis

Dennis, K. and Miles, R. S., 1979. Eubrachythoracid arthrodires with tubular rostral plates from Gogo, Western Australia. **J. Linn. Soc. (Zool.) 67:** 297-328.

Gardiner, B. G. and Miles, R. S., 1975. Devonian fishes of the Gogo Formation, Western Australia. **Colloques. Internat. Cent. Nat. Rech. Scient. 218:** 73-79.

Long, J. A., 1982. The history of fishes on the Australian continent. Pp. 53-86. in "The fossil vertebrate record of Australasia" ed by P. V. Rich and E. M. Thompson. Monash University Offset Printing Unit: Melbourne.

Culmacanthus stewarti

Long, J. A., 1982. The history of fishes on the Australian continent. Pp. 53-86 in "The fossil vertebrate record of Australasia" ed by P. V. Rich and E. M. Thompson. Monash University Offset Printing Unit: Melbourne.

Long, J. A., 1983a. New Bothriolepid fishes from the late Devonian of Victoria, Australia. **Palaeontology. 26:** 295-320.

Long, J. A., 1983b. A new diplacanthoid acanthodian from the late Devonian of Victoria. **Mem. Ass. Australia. Palaeontol.** No. 1

Long, J. A., Turner, S. and Kemp, A., 1982. Contributions to Australian fossil fish biostratigraphy. Pp. 119-44 in "The fossil vertebrate record of Australasia" ed by P. V. Rich and E. M. Thompson. Monash University Offset Printing Unit: Melbourne.

Neoceratodus gregoryi

White, E. I., 1925. Two new fossil species of **Epiceratodus** from South Australia. **Ann. Mag. Nat. Hist. 9(16):** 139-46.

Long, J. A., Turner, S. and Kemp, A., 1982. Contributions to Australian fossil fish biostratigraphy. Pp. 119-44 in "The fossil vertebrate record of Australasia" ed by P. V. Rich and E. M. Thompson. Monash University Offset Printing Unit: Melbourne.

Paracyclotosaurus davidi

Chernin, C. and Cruickshank, A. R. I., 1978. The myth of the bottom dwelling capitosaur amphibians. **Sth. Afr. J. Sci. 74:** 111-12.

Watson, D. M. S., 1958. A new labyrinthodont (**Paracyclotosaurus**) from the Upper Trias of New South Wales. **Bull. Br. Mus. Nat. Hist. 3:** 233-64.

Rhoetosaurus brownei

Longman, H. A., 1926. A giant dinosaur from Durham Downs, Queensland. **Mem. Qd. Mus. 8:** 183-94.

Longman, H. A., 1927. The giant dinosaur, **Rhoetosaurus brownei. Mem. Qd. Mus. 9:** 1-18.

Thulborn, R. A., 1982. Speeds and gaits of dinosaurs. **Paleogeography, Paleoclimatography, Paleoecology. 38:** 227-56.

Allosaurus

Molnar, R. E., Flannery, T. F. and Rich, T. H. V., 1981. An allosaurid theropod dinosaur from the early Cretaceous of Victoria, Australia. **Alcheringa 5:** 141-47.

Flannery, T. F. and Rich, T. H., 1981. Dinosaur digging in Victoria. **Aust. Nat. Hist. 20** (6): 195-8.

Romer, A. S., 1966. Vertebrate Paleontology. **Chicago Univ. Press:** Chicago.

Muttaburrasaurus langdoni

Bartholomai, A. and Molnar, R. E., 1981. **Muttaburrasaurus**, a new iguanodontid (Ornithischia: Ornithopoda) dinosaur from the Lower Cretaceous of Queensland. **Mem. Qd. Mus. 20:** 319-49.

Molnar, R. E., 1982. Australian Mesozoic reptiles. Pp. 170-225 in "The fossil vertebrate record of Australasia" ed by P. V. Rich and E. M. Thompson, Monash University Offset Printing Unit: Melbourne.

Kronosaurus queenslandicus

Longman, H. A., 1924. Some Queensland fossil vertebrates. **Mem. Qd. Mus. 8:** 16-28.

Molnar, R. E., 1983. Australian Mesozoic reptiles. Pp. 169-225 in "The fossil vertebrate record of Australasia" ed by P. V. Rich and E. M. Thompson. Monash University Offset Printing Unit: Melbourne.

Romer, A. S. and Lewis, A. D., 1959. A mounted skeleton of the giant plesiosaur **Kronosaurus. Breviora 112:** 1-15.

White, T. E., 1935. On the skull of **Kronosaurus queenslandicus** Longman. **Occ. Pap. Boston Soc. Nat. Hist. 8:** 219-28.

Pterosaurs

Molnar, R. E., 1982. Australian Mesozoic reptiles. Pp. 170-228 in "The fossil vertebrate record of Australasia" ed by P. V. Rich and E. M. Thompson. Monash University Offset Printing Unit: Melbourne.

Molnar, R. E. and Thulborn, R. A., 1980. First pterosaur from Australia. **Nature. 288:** 361-63.

Meiolania platyceps

Anderson, C., 1925. Notes on the extinct chelonian **Meiolania**, with a record of a new occurrence. **Rec. Aust. Mus. 14:** 223-42.

Gaffney, E. S., 1981. A review of the fossil turtles of Australia. **Amer. Mus. Novitates** No. 2720.

Owen, R., 1886. Description of fossil remains of two species of megalanian genus (**Meiolania**) from "Lord Howe's Island". **Phil. Trans. Roy. Soc.** London. **177:** 471-80.

Ritchie, A., 1978. An island sanctuary. **Hemisphere 22(4):** 2-7.

Megalania prisca

Hecht, M., 1975. The morphology and relationships of the largest known terrestrial lizard, **Megalania prisca** Owen, from the Pleistocene of Australia. **Proc. Roy. Soc. Vict., 87:** 239-50.

Rich, T. H., and Hall, B., 1979. Rebuilding a giant. **Aust. Nat. Hist., 19:** 310-14.

Wonambi naracoortensis

Smith, M. J. 1976. Small fossil vertebrates from Victoria Cave, Naracoorte, South Australia iv Reptiles. **Trans. R. Soc. S.Aust. 100:** 39-51.

Dromornis stirtoni

Rich, P. V., 1979. The Dromornithidae, a family of large, extinct ground birds endemic to Australia. **Bur. Min. Res. Bull. 184:** 196.

Rich, P. V., 1982. Where did Australia's birds come from? — A question complicated by a new tectonic theory of the Earth. **Labtalk. Occ. Pap. Earth Sci., Sci. Teachers Assoc. Victoria. 35:** 19-30.

Rich, P. V. and Van Tets, G. F., 1982. Fossil birds of Australia and New Guinea: their biogeographic, phylogenetic, and biostratigraphic input. Pp. 235-38 in "The vertebrate fossil record of Australasia". ed by P. V. Rich and E. M. Thompson. Monash University Offset Printing Unit.

Rich, P. V. and Belouet, C., 1983. The waifs and strays of the bird world; or the ratite problem visited one more time. In: Archer, M. and Clayton, G. (eds). "Vertebrate zoogeography and evolution in Australasia." Hesperion Press: Perth.

Obdurodon insignis

Archer, M., Plane, M. D. and Pledge, N., 1978. Additional evidence for interpreting the Miocene **Obdurodon insignis**, Woodburne and Tedford, 1975 to be a fossil platypus (Ornithorhynchidae: Monotremata) and a reconsideration of the status of **Ornithorhynchus agilis** De Vis, 1885. **Aust. Zool. 30:** 9-27.

Woodburne, M. O. and Tedford, R. H., 1975. The first Tertiary monotreme from Australia. **Amer. Mus. Novit. 2588:** 1-11.

Augee, M., 1983. The curious problem of monotreme zoogeography. In "Vertebrate zoogeography and evolution in Australasia" ed by M. Archer and G. Clayton. Hesperion Press: Perth.

Zaglossus

Murray, P., 1978 a. A Pleistocene spiny anteater from Tasmania (Monotremata: Tachyglossidae, Zaglossus). **Papers and Proceedings Roy. Soc. Tas. 112:** 39-69.

Murray, P., 1978 b. Late Cenozoic monotreme anteaters. **Aust. Zool. 20:** 29-55.

Pledge, N., 1980. Giant echidnas in South Australia. **S. Aust. Nat. 55:** 27-30.

Dunn, W. S., 1895. Notes on the occurrence of monotreme remains in the Pliocene of New South Wales. **Rec. Geol. Surv. New South Wales 4:** 113-26.

Thylacinus potens

Woodburne, M. O., 1967. The Alcoota fauna, central Australia: an integrated palaeontological and geological study. **Bur. Min. Resour. Aust. Bull. 87:** 1-187.

Archer, M., 1982. A review of Miocene thylacinids (Thylacinidae, Marsupialia), the phylogenetic position of the Thylacinidae and the problem of apriorisms in character analysis. Pp. 445-76 in "Carnivorous marsupials" ed by M. Archer. Roy. Zool. Soc. New South Wales: Sydney.

Dawson, L., 1982. Taxonomic status of fossil thylacines (**Thylacinus**, Thylacinidae, Marsupialia) from late Quaternary deposits in eastern Australia. Pp. 527-36 in "Carnivorous marsupials" ed by M. Archer, Roy. Zool. Soc. New South Wales: Sydney.

Archer, M. and Clayton, G., 1983. "Vertebrate zoogeography and evolution in Australasia." Hesperion Press: Perth.

Glaucodon ballaratensis

Stirton, R. A., 1957. Tertiary marsupials from Victoria, Australia. **Mem. Nat. Mus. Vict. 21:** 121-34.

Archer, M., 1982. Review of the dasyurid (Marsupialia) fossil record, integration of data bearing on phylogenetic interpretation, and suprageneric classification. Pp. 397-433 in "Carnivorous marsupials" ed by M. Archer. Roy. Zool. Soc. New South Wales: Sydney.

Archer, M., 1983. The origins and radiations of the Australian marsupial radiation. In "Vertebrate zoogeography and evolution in Australasia" ed by M. Archer and G. Clayton. Hesperion Press: Perth.

Thylacoleo carnifex

Broom, R., 1898. On the affinities and habits of **Thylacoleo. Proc. Linn. Soc.** NSW (2). **22:** 57-74.

Finch, M. E., 1971. **Thylacoleo**, marsupial lion or marsupial sloth? **Aust. Nat. Hist. 17:** 7-11.

Finch, M. E., 1982. The discovery and interpretation of **Thylacoleo carnifex** (Thylacoleonidae, Marsupialia) Pp. 537-551 in M. Archer (ed) Carnivorous Marsupials, Symp. R. Zool. Soc. NSW.

Gill, E. D., 1954. Ecology and distribution of the extinct giant marsupial, **Thylacoleo. Vict. Naturalist 71:** 18-35.

Owen, R., 1859. On the fossil mammals of Australia. Part I. Description of a mutilated skull of a large marsupial carnivore (**Thylacoleo carnifex** Owen), from a calcareous conglomerate stratum eighty miles S.W. of Melbourne, Victoria. **Phil. Trans. R. Soc. 149:** 309-22.

Turnbull, W. D., 1970. Mammalian Masticatory Apparatus. Fieldiana: **Geology 18** (2): 149-356.

Wells, R. T. and Nichol, B., 1977. On the manus and pes of **Thylacoleo carnifex** Owen (Marsupialia) **Trans. R. Soc. S. Aust. 101:** 139-46.

Woods, J. T., 1956. The skull of **Thylacoleo carnifex. Mem. Qld. Mus. 13:** 125-40.

Palorchestes azael

Bartholomai, A., 1978. The rostrum in **Palorchestes** Owen (Marsupialia, Diprotodontidae). Results of the Ray. E. Lemley Expeditions, Part 3. **Mem. Qd. Mus. 18:** 145-49.

Owen, R., 1874. On the fossil mammals of Australia. Part 9. **Phil. Trans. Roy. Soc. Lond. 164:** 783-803.

Owen, R., 1876. On the fossil mammals of Australia. Part 10. Family Macropodidae: Mandibular dentition and parts of the skeleton of **Palorchestes**; additional evidences of **Macropus titan, Sthenurus** and **Procoptodon. Phil. Trans. Roy. Soc. Lond. 166:** 197-226.

Woods, J. T., 1958. The extinct marsupial genus **Palorchestes** Owen. **Mem. Qd. Mus. 13:** 177-93.

Neohelos

Stirton, R. A., Tedford, R. H. and Woodburne, M. O., 1967. A new Tertiary formation and fauna from the Tirani Desert, South Australia. **Rec. S. Aust. Mus. 15:** 427-62.

Stirton, R. A., 1967. A diprotodontid from the Miocene Kutjamarpu fauna, South Australia. **Bull. Bur. Miner. Resour. Geol. Geophys. Aust. 85:** 45-51.

Tedford, R. H., 1867. Fossil mammal remains from the Tertiary Carl Creek Limestone, north-western Queensland. **Bull. Bur. Miner. Resour. Geol. Geophys. Aust. 92:** 217-37.

Stirton, R. H., Woodburne, M. D. and Plane, M. D., 1967. A phylogeny of the Tertiary Diprotodontidae and its significance in correlation. **Bull. Bur. Miner. Resour. Geol. Geophys. Aust. 85:** 149-60.

Clemens, W. A. and Plane, M. D., 1974. Mid-Tertiary Thylacoleonidae (Marsupialia, Mammalia). **J. Paleo. 48:** 652-60.

Zygomaturus trilobus

Scott, H. H., 1915. A monograph of **Nototherium tasmanicum.** Tasmanian Dept. Mines. Geological Survey Rec. No. 4: 1-46.

Whitley, G. P., 1966. Some early references to the extinct marsupial **Zygomaturus. Aust. Zool. 13:** 228-30.

Stirton, R. A., 1967., New species of **Zygomaturus** and additional observations on **Meniscolophus,** Pliocene Palankarinna fauna, South Australia. **Bull. Bur. Miner. Res. Geol. Geophys. Aust. 85:** 129-47.

Hope, J. H., 1982. Late Cainozoic vertebrate faunas and the development of aridity in Australia. Chapter 8 **in** "Evolution of the flora and fauna of arid Australia" ed by W. R. Barker and P. J. M. Greenslade. Peacock Publications: Fewville.

Euryzygoma dunense

Archer, M. and Wade M. D., 1976. Results of the Ray E. Lemley expedition. Part 1. The Allingham Formation and a new Pliocene vertebrate fauna from northern Queensland. **Mem. Qd. Mus. 17:** 379-97.

De Vis, C. W., 1887. On a supposed new species of **Nototherium. Proc. Linn. Soc. N.S.W. 2:** 1065-70.

Emerson, K., 1964. An extinct marsupial of the Chinchilla district. **Wildlife in Australia. 1:** 15.

Longman, H. A., 1921. A new genus of fossil marsupials. **Mem. Qd. Mus. 7:** 65-80.

Stirton, R. A., Woodburne, M. W. and Plane, M. D., 1967. Tertiary Diprotodontidae from Australia and New Guinea. **Bull. Bur. Miner. Res. Geol. Geophys. Aust. 85.**

Diprotodon optatum

Rich, T. H. V., 1981. Australia's largest marsupial, **Diprotodon:** its ancestry, palaeobiology and extinction. Sci. Teachers Assoc. Victoria, Labtalk Ser. **34:** 21-28.

Tedford, R. H., 1973. The Diprotodons of Lake Callabonna. **Aust. Nat. Hist. 17:** 349-54.

Phascolonus gigas

Stirling, E. C., 1913. Fossil remains of Lake Callabonna. iv (2) On the identity of **Phascolomys (Phascolonus) gigas,** Owen, and **Sceparnodon ramsayi,** Owen with a description of some of its remains. **Mem. Roy. Soc. S. Aust. 1:** 111-78.

Ride, W. D. L., 1967. On **Sceparnodon ramsayi** Owen, 1884: the selection of a lectotype, the clarification of its type locality, and on its identity with **Phascolonus gigas** (Owen, 1858). **Rec. S. Aust. Mus. 15:** 419-25.

Wabularoo naughtoni

Archer, M., 1979. **Wabularoo naughtoni** gen. et sp. nov., an enigmatic kangaroo (Marsupialia) from the middle Tertiary Carl Creek Limestone of northwestern Queensland. Results of the Ray E. Lemley Expeditions, part 4. **Mem. Qd. Mus. 19:** 299-307.

Flannery, T. F., Archer, M. and Plane, M. D., 1983. Miocene kangaroos (Macropodoidea: Marsupialia) from three localities in northern Australia. **Bull. Bur. Miner. Res. Aust. Geol. Geophys. 7:** 287-302.

Flannery, T. F., 1983. The kangaroos: 15 million years of bounders. In "Vertebrate zoogeography and evolution in Australasia". ed by M. Archer and G. Clayton. Hesperion Press, Perth.

Bohra paulae

Flannery, T. F. and Szalay, F., 1982 **Bohra paulae,** a new giant fossil tree-kangaroo (Marsupialia; Macropodidae) from New South Wales, Australia. **Aust. Mammal. 5:** 83-95.

Flannery, T. F., 1983. The kangaroos: 15 million years of bounders. In "Vertebrate zoogeography and evolution in Australasia". ed by M. Archer and G. Clayton. Hesperion Press: Perth.

Troposodon kenti

Bartholomai, A., 1967. **Troposodon,** a new genus of fossil Macropodinae (Marsupialia). **Mem. Qd. Mus. 15:** 21-33.

Campbell, C., 1973. A new species of **Troposodon** Bartholomai; from the early Pleistocene Kanunka fauna, South Australia (Macropodinae: Marsupialia). **Rec. S. Aust. Mus. 16:** 1-18.

Flannery, T. F., 1983. The kangaroos: 15 million years of bounders. In "Vertebrate zoogeography and evolution in Australia". ed by M. Archer and G. Clayton. Hesperion Press: Perth.

Flannery, T. F., 1983. Review of the subfamily Sthenurinae (Marsupialia) and the relationships of the species of **Troposodon** and **Lagostrophus. Aust. Mammal. 6:** in press.

Procoptodon pusio

Stirton, R. D. and Marcus, L. F., 1966. Generic and specific diagnoses in the giant macropodid genus **Procoptodon. Rec. Aust. Mus. 26:** 349-59.

Bartholomai, A., 1970. The extinct genus **Procoptodon** Owen (Marsupialia: Macropodidae) in Queensland. **Mem. Qd. Mus. 15:** 213-33.

Dolphins

Fordyce, E., 1983. Rhabdosteid dolphins (Mammalia: Cetacea) from the middle Miocene, Lake Frome area, South Australia. **Alcheringa. 7:** 27-40.

Tedford, R. H., Archer, M., Bartholomai, A., Plane, M., Pledge, N. S., Rich, T. H., Rich, P. V. and Wells, R. T., 1977. The discovery of Miocene vertebrates, Lake Frome area, South Australia. **Bull. Bur. Miner. Res. Aust. Geol. Geophys. 2:** 53-57.

An Ancient Australian bat

Hall, L. S. and Richards, G. C., 1979. "Bats of eastern Australia." Queensland Museum: Brisbane.

Vaughan, T. A., 1976. Nocturnal behaviour of the African false vampire bat (**Cardioderma cor**). **J. Mammal. 57:** 227-48.

Petersen, R., 1964. "Silently by night". Longmans, Green and Co. Ltd: London.

Hand, S. J., 1983. Bat beginnings and biogeography: a southern perspective. In "Vertebrate zoogeography and evolution in Australasia", ed by M. Archer and G. Clayton. Hesperion Press: Perth.